Monte Carlo-based technique for uncertainty calculations

Raul R. Cordero

Monte Carlo-based technique for uncertainty calculations

and its applications to UV spectroradiometry

Südwestdeutscher Verlag für Hochschulschriften

Impressum/Imprint (nur für Deutschland/only for Germany)
Bibliografische Information der Deutschen Nationalbibliothek: Die Deutsche Nationalbibliothek verzeichnet diese Publikation in der Deutschen Nationalbibliografie; detaillierte bibliografische Daten sind im Internet über http://dnb.d-nb.de abrufbar.
Alle in diesem Buch genannten Marken und Produktnamen unterliegen warenzeichen-, marken- oder patentrechtlichem Schutz bzw. sind Warenzeichen oder eingetragene Warenzeichen der jeweiligen Inhaber. Die Wiedergabe von Marken, Produktnamen, Gebrauchsnamen, Handelsnamen, Warenbezeichnungen u.s.w. in diesem Werk berechtigt auch ohne besondere Kennzeichnung nicht zu der Annahme, dass solche Namen im Sinne der Warenzeichen- und Markenschutzgesetzgebung als frei zu betrachten wären und daher von jedermann benutzt werden dürften.

Coverbild: www.ingimage.com

Verlag: Südwestdeutscher Verlag für Hochschulschriften GmbH & Co. KG
Heinrich-Böcking-Str. 6-8, 66121 Saarbrücken, Deutschland
Telefon +49 681 37 20 271-1, Telefax +49 681 37 20 271-0
Email: info@svh-verlag.de

Approved by: Hannover: Leibniz Universität Hannover, Diss., 2012

Herstellung in Deutschland (siehe letzte Seite)
ISBN: 978-3-8381-3438-3

Imprint (only for USA, GB)
Bibliographic information published by the Deutsche Nationalbibliothek: The Deutsche Nationalbibliothek lists this publication in the Deutsche Nationalbibliografie; detailed bibliographic data are available in the Internet at http://dnb.d-nb.de.
Any brand names and product names mentioned in this book are subject to trademark, brand or patent protection and are trademarks or registered trademarks of their respective holders. The use of brand names, product names, common names, trade names, product descriptions etc. even without a particular marking in this works is in no way to be construed to mean that such names may be regarded as unrestricted in respect of trademark and brand protection legislation and could thus be used by anyone.

Cover image: www.ingimage.com

Publisher: Südwestdeutscher Verlag für Hochschulschriften GmbH & Co. KG
Heinrich-Böcking-Str. 6-8, 66121 Saarbrücken, Germany
Phone +49 681 37 20 271-1, Fax +49 681 37 20 271-0
Email: info@svh-verlag.de

Printed in the U.S.A.
Printed in the U.K. by (see last page)
ISBN: 978-3-8381-3438-3

Preface

A Monte Carlo-based technique has been developed to assess the uncertainty of surface UV data. This new method allows evaluating the uncertainty of both modeled and measured spectra, contributing in turn to ensure their quality. Quality-ensured UV series are required for an improved understanding of the global UV climate.

1-D radiative transfer models allow *calculating* surface UV spectra under cloudless conditions. The uncertainty of modeled UV spectra is due to uncertainties in the input quantities needed to run 1-D models: ozone, aerosol properties, extraterrestrial spectrum, albedo, etc. The effects of uncertainties associated with these input quantities are strongly nonlinear and therefore cannot be described by using conventional uncertainty propagation techniques.

Spectroradiometers allow *measuring* surface UV spectra regardless of the cloud conditions. In this case, uncertainty arises not only when measuring in field, but also when calibrating in the lab. Assessing the effect of correlations between measurements and calibrations may be difficult when applying conventional techniques.

The new Monte Carlo-based technique allowed overcoming those difficulties and fully addressing the uncertainty evaluation problem when assessing surface UV data. On the one hand, the performance of radiative transfer models was found to be significantly dependent on the amount of aerosols above the measuring site; if quality-ensured inputs are available to feed the model, under cloudless conditions the UV-B uncertainties are up to 18% for clean sites, and up to 40% for sites with large aerosol load (see section 4). On the other hand, the uncertainty of measurements carried out by state-of-the-art spectroradiometers was found to be significantly lower:

1

about 6% in the UV-A and up to 9% in the UV-B (see section 5).

The potential applications of the Monte Carlo-based technique were further explored; it was realized that the proposed technique is able to describe the uncertainty propagation through any process of exploiting experimental data. For example, the proposed approach was used to evaluate the uncertainties associated with the UV index, the ozone column, and the aerosol optical depth (AOD), all retrieved from quality-controlled UV spectra. It was found that UV indexes with uncertainties of about 6-8% could be computed by integrating global UV irradiances (see section 6). Uncertainties of about 8% for the ozone column, and of about 22% for AOD retrievals, were found when exploiting the direct UV irradiance (see section 7).

The new Monte Carlo-based technique is general, meaning that it allows comprehensively describing the uncertainty propagation through any measuring process or any retrieving scheme. Therefore, it has the potential to become a useful tool for exploiting spectral UV measurements and for ensuring their quality. The proposed technique agrees with recommendations of the ISO Guide to the Expression of Uncertainty in Measurement.

Keywords: Monte Carlo, Uncertainty, UV radiation

Vorwort

Eine Monte Carlo basierte Technik wurde entwickelt, um die Unsicherheit der spektralen Bestrahlungsstärke (sowohl durch Strahlungstransfermodelle, als auch durch Spektralradiometer) abzuschätzen.

Bei Strahlungstransfermodellen (siehe Kapitel 4) entsteht die Unsicherheit der UV Strahlung durch die Unsicherheit der Messungen, welche die Modelle erfordern: Ozonwerte, Aerosolkonzentration, extraterrestrisches Spektrum, und Albedo. Diese Eingabeparameter unterliegen gewissen Unsicherheiten; die Auswirkungen dieser Unsicherheiten sind nicht linear, daher ist es nicht möglich, sie durch Anwendung konventioneller Techniken darzustellen.

Bei Spektralradiometern (siehe Kapitel 5) entsteht die Unsicherheit nicht nur bei der Messung, sondern auch bei Kalibrierung. Darüber hinaus wird die Anwendung konventioneller Techniken durch die Korrelation zwischen der Unsicherheit durch Messungen und Kalibrierungen erschwert.

Die in dieser Arbeit beschriebene Monte Carlo basierte Technik ermöglicht es, diese Schwierigkeiten in der Unsicherheitsabschätzung zu bewältigen und dem Problem der Unsicherheitsfortpflanzung der UV spektrale Bestrahlungsstärke zu begegnen.

Die Aerosolkonzentration beeinflusst die Unsicherheit des Strahlungstransfermodells; die Unsicherheiten der Bestrahlungsstärke im UVB-Bereich liegen bei wolkenlosem Himmel zwischen 18% (für niedrige Aerosolkonzentration) und 40% (für höhere Aerosolkonzentration). Messungen durch Spektralradiometer der Bestrahlungsstärke im UVB-Bereich haben dem gegenüber eine Unsicherheit von ca. 9%.

UV spektrale Bestrahlungsstärke kann einerseits zur UV-Index Berechnung ausgewertet werden, sowie zur Gewinnung von Ozonwerten und Aerosol Optische Dicke (AOD). Obwohl verschiedene Techniken existieren, geben sie keine Informationen zur Erfassung von Unsicherheiten.

Wie in den Kapiteln 6 und 7 genauer erläutert, ermöglicht die Monte Carlo basierte Technik, die Unsicherheit der UV-Index, Ozonwerte, und AOD abzuschätzen. Man fand heraus, dass die Unsicherheit des UV-Indexes bei einem Sonnenzenitwinkel weniger als 30° ca. 6-8% betrug. Zur Gewinnung der Ozonwerte und AOD bei Messungen der direkten Bestrahlungsstärke wurden Unsicherheiten ca. 8% bzw. von ca. 22% geschätzt .

Die Monte Carlo basierte Technik trägt zur Lösung des Problems der Unsicherheitsforpflanzung bei und kann somit in naher Zukunft ein nützliches Mittel zur Gewährleistung der Qualität der spektralen UV Messungen werden. Die Technik ist allgemein gehalten und stimmt mit den Empfehlungen des ISO Guide to the Expression of Uncertainty in Measurement überein.

Schlagwörter: Unsicherheit, Monte Carlo, UV Strahlung.

Table of Contents

1. Introduction

1.1. UV radiation

The electromagnetic radiation that the earth receives from the sun provides a tremendous amount of energy every day. Its maximum is in the part of the spectrum known as shortwave solar radiation (300-2500 nm wavelength, see Gueymard, 2004), which consists of infrared, visible and ultraviolet radiation. The ultraviolet (UV) solar spectrum is in turn divided into three regions: the UV-C (< 280 nm wavelength), which is strongly absorbed by the atmosphere and therefore it is undetectable by ground-based measurements, UV-A (315-400 nm) and UV-B (280-315 nm).

Increases in surface UV radiation mostly resulting from the well-known depletion in total column ozone is an important environmental concern; the UV radiation is known to have adverse effects on the biosphere including terrestrial and aquatic ecosystems as well as on public health; several plants react to increased UV radiation with reduced growth or diminished photosynthetic activity (Tevini, 1993; Slaper et al, 1996). Although UV radiation is necessary for the synthesis of vitamin D in the human skin (Autier and Gandini, 2007; Holick, 2008; Bogh et al, 2011), exposure to UV radiation is associated with skin cancer, accelerated ageing of the skin, cataract and other eye diseases. It may also affect people's ability to resist infectious diseases and compromise the effectiveness of vaccination programmes.

As international standard measure of the UV level, which can lead to an erythemal or sunburning response in humans is used the UV index. This is evaluated by calculating the integral in the range 250-400 nm of the spectral UV irradiance weighted by using the so-called McKinlay-Diffey Erythema action spectrum (McKinlay and Diffey, 1987). While some of the adverse effects of the UV irradiance may be strictly proportional to cumulative UV dose, others may relate to the frequency of extreme

7

UV-B events (WMO, 1997). Therefore an improved understanding of the global UV climate, including variability and trends in surface UV irradiance, has become of great interest.

1.2. Surface UV radiation

The sun's total and spectral radiation reaching ground is strongly modified by absorption and scattering processes, such as scattering and reflection by clouds, Rayleigh scattering by air molecules and absorption by atmospheric ozone, water vapor and CO_2 (Jacovides et al, 2000; Zerefos, 2002).

The effect of aerosols on UV irradiance is complex, owing to the variety of composition of aerosol material that may be present in the atmosphere, as well as its varying distribution (Jacovides et al, 2000). The aerosol assessment is difficult because of the difficulties involved in the measurements the parameters that characterize its effect. However, as shown below (see section 7), reliable evaluations of the aerosol optical depth (AOD) can be obtained from the ground-based measurements of the direct irradiance.

Clouds are responsible for a great deal of the observed irradiance variability; cloud cover is an important factor in reducing the surface UV-B radiation (Foyo-Moreno et al, 2003; Krzyscin et al, 2003). However, cloud effects on UV irradiance are difficult to quantify. In practice, the parameters needed to assess the cloud effect are rarely available and even if they are, the complexities of cloud geometry (with inhomogeneous or broken cloud fields) as well as heterogeneous terrains would demand the use of complex 3-D models, which require significant computational time and are therefore are restricted to a subset of conditions. Major difficulty with 3-D models is the fact that the Sun can be unobscured even for large cloud fractions, or obscured even for small cloud fractions, makes the quantification of cloud effects problematic (e.g. Davis and Marschak, 2010; Kato et al, 2009; Zinner et al, 2006; Matthijsen et al, 2000; Udelhofen et al, 1999).

1.3. Ozone depletion

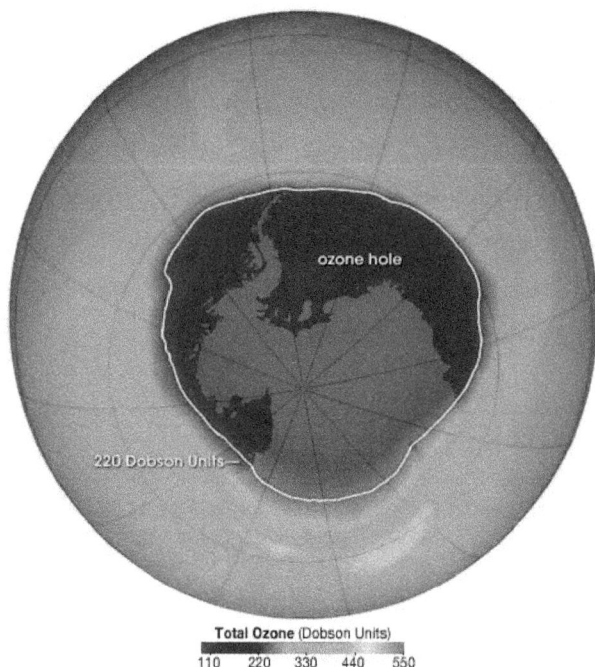

Figure 1.1[1]. Total ozone map over the Antarctica on 10.09.2010 based on data from OMI on board the AURA satellite. The data are processed and mapped by NASA..

Ultraviolet radiation at Earth's surface increases as the amount of overhead total ozone decreases because ozone absorbs ultraviolet radiation from the Sun. Ozone is the most important absorber of UV-B radiation (Zerefos, 2002) leading to differences

---

[1] Plot quoted from (http://ozonewatch.gsfc.nasa.gov/)

9

of several orders of magnitude over a relatively short wavelength range (290 to 315 nm).

The ozone layer has been depleted gradually since 1980 and is about an average of 3% lower over the globe (WMO, 2011). The depletion, which exceeds the natural variations of the ozone layer, appeared to be due to the action of the so-called halogen source gases (also known as ozone-depleting gases) such as the chlorofluorocarbons (CFCs), hydrochlorofluorocarbons (HCFCs), and halons. These human-produced gases, associated with certain industrial processes, eventually reach the stratosphere where they are broken apart to release Ozone-depleting atoms. The production and consumption of the main halogen source gases by human activities are regulated worldwide under the Montreal Protocol (WMO, 2011).

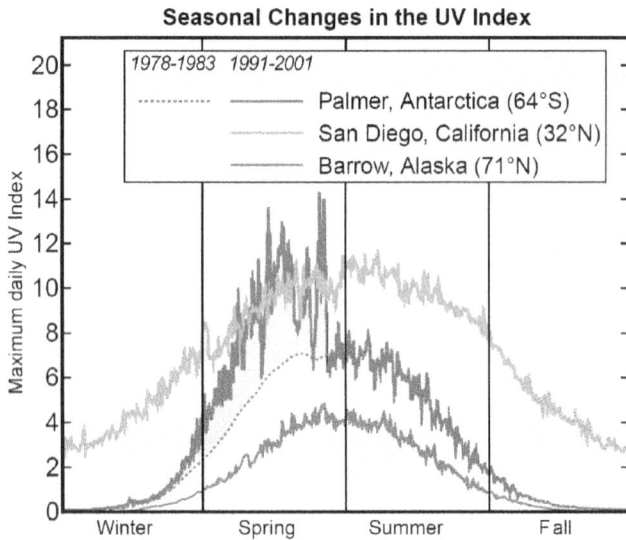

Figure 1.2[2]. Seasonal changes in the UV index. Plot quoted from (WMO, 2007).

---

[2] Plot quoted from (WMO, 2007)

The ozone depletion is very small near the equator and increases with latitude toward the poles; the phenomenon of the "Antarctic ozone hole" (a large average depletion on that polar region) is primarily a result of the late winter/spring ozone destruction that occurs there annually because of the very cold temperatures of the Antarctic stratosphere. The size of the "Antarctic ozone hole" is usually defined by the spatial extent of the polar vortex (see figure 1.1). During October and November, low stratospheric ozone concentrations are centered at South Pole and show a sharp boundary at about 65°S latitude. However, the instability of the ozone hole can move packages of ozone-depleted air masses towards lower latitudes (Uchino et al, 1999).

Similar springtime ozone depletion occurs at March also at Northern Polar latitudes. Although there, due to the slight higher temperatures, it is not as intense as at the Antarctic latitudes (see Solomon et al, 2007), unprecedented Arctic ozone loss has been recently reported (Manney GL, 2011).

In the Antarctic, ozone depletion has been the dominant factor for increases in UV irradiance (Bernhard et al, 2010). The effect of ozone hole on the UV index is demonstrated by comparing the Palmer and San Diego data in the figure 1.2. Normal values estimated for Palmer are shown for the 1978-1983 period before the "ozone hole" occurred (see dotted line). In the last decade (1991-2001), Antarctic ozone depletion has led to an increase in the maximum UV Index value at Palmer throughout spring (see yellow shaded region). Values at Palmer (see Bernhard et al, 2005, for details) now sometimes equal or exceed those measured in San Diego, which is located at a lower latitude. The World Health Organization considers that values of UV index greater than 11 stand for extreme risk of harm from unprotected sun exposure.

As a result of the Montreal Protocol, the total abundance of ozone-depleting gases in the atmosphere has begun to decrease in recent years. It is expected that the decrease will continue throughout the 21st century. Some individual gases such as halons and hydrochlorofluorocarbons (HCFCs) are still increasing in the atmosphere, but will begin to decrease in the next decades if compliance with the Montreal Protocol continues. By midcentury, the effective abundance of ozone-depleting gases should fall to values present before the Antarctic "ozone hole" began to form in the early 1980s (WMO, 2011).

1.4. Climate change link

The radiative balance of Earth's atmosphere have changed by the abundances of the so-called "greenhouse gases". These gases result in radiative forcings, which can lead to climate change. The largest radiative forcings come from carbon dioxide, followed by methane, tropospheric ozone (which results from pollution associated with human activities), the halogen-containing gases, and nitrous oxide. All these forcings are positive, which leads to a warming of Earth's surface. Ozone-depleting gases (such as the chlorofluorocarbons (CFCs), hydrochlorofluorocarbons (HCFCs), and halons) also contribute to climate change because the stratospheric ozone depletion stands for a small negative forcing, which leads to cooling of Earth's surface (WMO, 2011).

Certain changes in Earth's climate could affect the future of the ozone layer and in turn the surface UV radiation. Stratospheric ozone is influenced by changes in temperatures and winds in the stratosphere. For example, low temperatures and strong polar winds both affect the extent and severity of winter polar ozone depletion. While Earth's surface is expected to warm in response to the positive radiative forcing from carbon dioxide increases, the stratosphere is expected to cool.

Further rise greenhouse gas emissions and further ozone depletion as a result of stratospheric cooling, can lead to drastic changes in the solar radiation climate on earth (WMO, 2011). The major parameters determining the spectral UV irradiance (clouds, total ozone column, aerosols and albedo) are likely to change significantly in their absolute amount, in their qualitative structure or in their temporal pattern. As a consequence, significant changes of the UV climate on earth are expected.

2. State-of-the-Art

2.1. Satellite estimates

The surface UV irradiance can be calculated from satellite measurements and by solving the equation of radiative transfer that governs the transfer of radiant energy in the atmosphere. However, these methods are indirect and necessarily involve assumptions about the spectral characteristics of clouds, aerosol attenuation and surface reflectivity; hence, they are less accurate than ground-based measurements.

The uncertainties of model results, on which all satellite-derived products are based (Martin et al, 2001), are high when clouds are considered. The derivation of UV irradiance from satellite instruments is problematic because they use backscattered ultraviolet radiation for the retrieval. Detailed studies have demonstrated that these satellite-based methods seriously underestimate UV irradiances in the northern hemisphere, where satellite-derived UV irradiance can sometimes exceed ground-based measurements by more than 40% (see McKenzie et al, 2001). Another approach is the derivation of UV irradiance by using geostationary satellites in combination with polar orbiting satellites (Verdebout, 2004a; Verdebout, 2004b; Wuttke et al, 2003). The deviation of these satellite products with respect to the ground-based measurements is about 10% smaller.

Recent validation of satellite estimates have been focused on OMI data (gathered by the Ozone Monitoring Instrument (OMI) aboard EOS Aura satellite of the NASA), which has been compared to ground-based measurements (Tanskanen et al, 2006; Tanskanen et al, 2007a; Weihs et al, 2008; Kazadzis et al, 2009a; Kazadzis et al, 2009b; Ialongo et al, 2008; Ialongo et al, 2009; Buchard et al, 2008). All found that the OMI-derived UV estimates are biased high, particularly at polluted sites.

Differences with ground-based data are often linked with the limited spatial resolution; satellite measurements represent a much larger region (OMI minimum pixel at nadir 13×24 km^2) than ground-based measurements. Ground-based measurement of UV erythemal dose at various sites within one OMI satellite pixel showed deviations of ±5% under cloud-free conditions, or 20% when including urban areas (Weihs et al, 2008; Kazadzis et al, 2009b). For partly cloudy conditions and overcast conditions the discrepancy of instantaneous values between the stations can exceed 200%. If 3-hourly averages are considered, the agreement is better than 20% within a distance of 10 km (Weihs et al, 2008). This spatial discrepancy can explain much of the differences between ground-based and satellite data but of course cannot explain a strong systematic bias.

The OMI overestimation of ground-based UV measurements may be partly explained by the lack of sensitivity of satellite instruments to the boundary layer (Weihs et al, 2008; Kazadzis et al, 2009a; Buchard et al, 2008; Ialongo et al, 2008).

However, over snow-covered surfaces the OMI-derived daily dose is generally lower than the ground-based measurement because the OMI surface UV algorithm uses climatological surface albedo that may then be lower than the actual effective surface albedo (Tanskanen et al, 2007b; WMO, 2011). Part of the problem is that a portion of the observed reflectivity may be incorrectly interpreted as cloud cover, which reduces the estimated irradiance. All-conditions data and snow-free data have been compared separately to evaluate the albedo effect. For example, a comparison by Buchard et al, (2008) demonstrated that OMI overestimates erythemal daily doses by 14% for days without snow on the surface and only by 8% if days with snow are included in the comparison.

Moreover, it is worth to highlight the elevated incertitude associated with nadir-viewing satellite measurements under high SZA. This is a condition that frequently occurs when satellites record data at high latitudes close to winter periods. This SZA dependence of a nadir-viewing satellite instrument is coming from the increasing altitude of the sunlight scattering layer at increasing SZA, not properly taken into account in the retrieval (e.g. Hendrick et al, 2011). This usually implies a seasonal dependence more evident when satellite readings are compared with ground-based information.

2.2. Modeled UV data

Radiative transfer models (Gary et al, 1999) allow calculating the surface UV irradiance from some set of measured *input quantities* linked with the surface reflectivity, the solar zenith angle, the ozone column as well as the spectral characteristics of clouds and aerosols. The characterization of the three-dimensional structure and the inhomogeneity of clouds requires 3-D Monte Carlo models (Zinner et al, 2006; Kato et al, 2009; Davis and Marschak, 2010). These models allows in addition to three-dimensional atmospheres, the realistic treatment of inhomogeneous surface albedo and topography, (see Kylling et al, 2000; Mayer, 1999; Mayer, 2000; Smolskaia, 2001; Smolskaia et al, 2006).

Despite recent progress, the problems due to the characterization of the cloud effect complicate the assessment of UV irradiance by using both radiative transfer models and satellite measurements and promote ground-based measurements. However, under cloudless sky conditions, 1-D model outcomes have shown to be very useful in order to make UV reconstructions (Junk et al, 2007; Janouch, 2007) and to check the consistency of long-term ground-based measurements (Bernhard et al, 2005).

The quality of model outcomes can be checked through systematic comparison of spectral UV measurements and spectral UV calculations; in the case of 1-D models, comparisons under cloudless conditions have been already carried out (Badosa et al,

2007; Satheesh et al, 2006; Weihs et al, 1999; Weihs et al, 1997a; Mayer et al, 1997). However, the uncertainty analysis of radiative models requires relating the uncertainty of the *output quantity* (the irradiance) to the uncertainties of the input quantities through an adequate propagation technique. This means that the uncertainty assessment can be performed independently of the comparisons and therefore, the uncertainty of ground-based measurements and the uncertainty of model outcomes can be separately evaluated.

Uncertainty analysis of the surface irradiance calculated by using 1-D radiative transfer models (Weihs and Webb, 1997b; Schwander et al, 1997) has been already reported. However the uncertainty propagation techniques applied in these works were unable to account for the nonlinear effects on the irradiance due to the contributions of some uncertainty sources.

In section 4, the Monte Carlo-based uncertainty propagation technique proposed in section 3 is used to evaluate the uncertainties of the spectral UV irradiance rendered by a 1-D radiative transfer model. This technique allowed expressing the uncertainty of the output quantities (the irradiances) in terms of the standard uncertainties of the input quantities (ozone column, albedo, aerosol properties, etc). As an example, the UVSPEC model (Mayer and Kylling, 2005) was used to calculate the global UV irradiances corresponding to two scenarios both under cloudless sky conditions: polluted and unpolluted.

2.3. UV irradiance measurements

Understanding the UV climate requires assessing the variability and detecting trends in surface UV irradiance. However, the problems due to the characterization of the cloud effects complicate the evaluation of surface UV irradiance by using both radiative transfer models and satellite measurements and promotes ground-based measurements.

The measurement of solar spectral UV irradiance has historically been a difficult task. The fact that irradiance varies by many orders of magnitude over a relatively short wavelength range (290 to 315 nm) requires instruments with wide dynamic range and low stray light levels. Also the long-term stability of UV instruments and their absolute calibration standards are still difficult to maintain. Consequently, good quality routine spectral measurements did not start until the late 1980s and these longer records are worldwide few in number.

The spectrally resolved UV irradiance can be nowadays efficiently measured by using double monochromators-based spectroradiometers. These *scanning* instruments have become the reference instruments to measure spectral UV irradiance. For trend detection, the Network for the Detection of Atmospheric Composition Change (NDACC) (former Network for the Detection of Stratospheric Change (NDSC)) and the World Meteorological Organization (WMO) have defined a set of strict specifications (Wuttke et al, 2006; McKenzie et al, 1997; Seckmeyer et al, 2001); attending to the influence of the uncertainty sources affecting ground-based measurements (Bernhard and Seckmeyer, 1999; Bais, 1997), offset suppression, noise minimization, stray light counteraction, radiometric stability are among of the required specifications. The latter is the most important requirement for trend detection; it can be determined by repeated checks against standard lamps. Some instruments, such as well-maintained Brewer double spectroradiometers, are known to be very stable against standard lamps and can produce radiance stability near 1% (Cede et al, 2006).

Global UV climate assessment requires comparing ground-based UV measurements carried out at different geographical locations; significant hemispherical differences

18

are expected (Seckmeyer et al, 2008; Seckmeyer and McKenzie, 1992). However, quality-controlled UV data are being obtained mostly from UV measuring stations in the Northern Hemisphere. There are few stations in South America (Argentina, Chile, and Brazil), New Zealand, Australia, and Africa.

The solution for the current lack of southern hemisphere spectral UV monitoring stations in the existing World Meteorological Organization (WMO) network, seems to be linked with the development of low cost array spectroradiometers. Instead of *scanning* the spectrum, in array instruments the spectrum is directly imaged on to a Charge Coupled Device (CCD) array (Jaekel et al, 2007; Ansko et al, 2008; Ylianttila et al, 2005; Coleman et al, 2008; Kouremeti et al, 2008). Although CCD arrays are successfully being used to measure visible radiation, UV spectral measurements carried by these instruments are strongly affected by stray light (mostly originated in the visible light) that tends to mask the UV irradiances and leads to biased spectra (Zong et al, 2006; Kreuter and Blumthaler, 2009; Jaekel et al, 2007). Either theoretically robust (Zong et al, 2006) or experimental-based (Kreuter and Blumthaler, 2009; Jaekel et al, 2007; Riechelmann 2008) the stray light corrections have allowed improving the array accuracy (tested by intercomparisons). However, the problem of the uncertainty evaluation of irradiances rendered by array instruments has not been comprehensibly addressed because of the difficulties involved in the uncertainty propagation.

2.4. Double monochromator-based spectroradiometers

The quality of measurements carried out by NDACC-certified instruments has been validated by the systematic comparison of spectral UV measurements (under cloudless sky conditions) with spectral UV calculations (Badosa et al, 2007; Satheesh et al, 2006; Weihs et al, 1999; Weihs et al, 1997; Mayer et al, 1997), and by intercomparison campaigns that involved several instruments (Gröbner et al, 2006; Bais et al, 2001; Gröbner et al, 2000).

Figure 2.1 shows some spectral UV irradiance measurements carried out by using a spectroradiometer system. Assessing the agreement when comparing measurements requires the prior uncertainty evaluation. The latter in turn requires assessing the combined influence of the involved uncertainty sources.

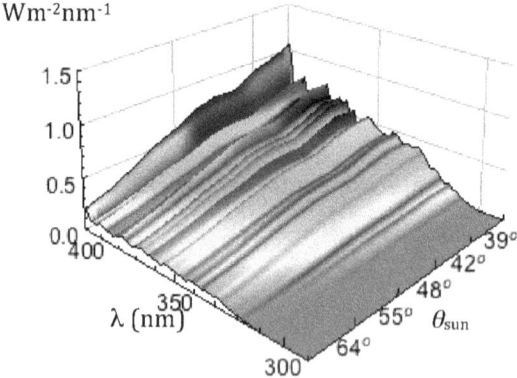

Figure 2.1: Ground-based UV irradiance measurements

However, the problem of the uncertainty evaluation of irradiances obtained by the NDACC instruments has not been comprehensibly addressed because of the difficulties involved in the uncertainty propagation. The irradiance evaluation from the experimental data rendered by NDACC spectroradiometers requires using information obtained during some prior adjustments (such as the absolute calibration and the wavelength alignment). This means that eventual errors in the prior calibrations affect also the irradiance measurements. These uncertainty sources can lead to nonlinear effects on the irradiances that cannot be fully described by applying the conventional Law of Propagation of Uncertainties (LPU) (ISO, 1993).

In section 5, it is carried out an uncertainty analysis of the spectral irradiances measured by using the spectroradiometer system of the Leibniz Universität Hannover (Institut für Meterologie und Klimatologie, IMUK), a NDACC-certified mobile instrument, in what follows referred to as IMUK spectroradiometer. The effects of uncertainties originated in the prior adjustments (absolute calibration and wavelength alignment) were explicitly considered in the uncertainty evaluation. Because the influences of these uncertainty sources on the spectral irradiance are expected to be nonlinear, a new Monte Carlo-based uncertainty propagation technique (see section 3) was applied.

The quality-ensured surface spectral UV radiation data rendered by fully characterized instruments have different applications. A couple of them were addressed: the computation of the UV index, and the retrieval of the aeorosol parameters from ground-based measurements of the spectral direct UV irradiance.

2.5. UV index

The need to integrate empirical data is of frequent occurrence in experimental activities. These experimental data are generally rendered by a measuring instrument as a two-dimensional set of points. The numeric integration of these points involves first constructing an Interpolating Function to approximate the underlying function that produced the data; then, the Interpolating Function is integrated to obtain the result.

Computing the UV index requires integrating the biologically-weighted surface ultraviolet irradiance. The latter is rendered by a spectroradiometer as a two dimensional set of experimental points. Since these experimental points are somehow uncertain, Interpolating Functions will also be uncertain, and so will be integrals computed from them.

In section 6, the problem of the uncertainty propagation in the computation of the UV index from the experimental data rendered by spectroradiometer systems is addressed. A new procedure based on a Monte Carlo simulation (see section 3) was used to evaluate the uncertainty associated with the UV index. This is evaluated by calculating the integral in the range 250-400 nm of the spectral ultra-violet irradiance weighted by using the McKinlay-Diffey Erythema action spectrum (McKinlay and Diffey, 1987). The latter describes the relative effectiveness of energy at different wavelengths in producing a particular biological response.

2.6. Aerosol Retrieval

Aerosols lead to attenuation in the radiant energy that is normally characterized by evaluating the aerosol optical depth (AOD). Several methods have been proposed in order to retrieve both the ozone column and the AOD (and in turn the Angström parameters) from spectral measurements of direct UV irradiance. Some of them (Cachorro et al, 2002; Mayer and Seckmeyer, 1998) imply sequentially removing from the total atmospheric optical depth, the contributions due to the Rayleigh scattering, the aerosol extinction and the ozone absorption. After removing these contributions, the values of the Angström parameters and the ozone column that lead to a minimal residual are considered to be the corresponding best estimates. A related method involves comparing the measured spectrum with model calculations (Huber et al, 1995). In this latter case, the best estimates of both the ozone column and the AOD are considered to be those values that lead to the best match.

Although these techniques are well established, they do not consider the uncertainties associated with the involved experimental data and are unable to describe the uncertainty propagation through the retrieving process.

In section 7, a new Monte Carlo-based exploiting approach (see section 3) is applied. This implied sequentially comparing the ground-based measurements with a large number of spectra, each of them calculated by using the UVSPEC radiative transfer model (Mayer and Kylling, 2005) fed with randomly generated values of *the aerosol parameters*. Some of the generated values led to a calculated spectrum that matched reasonably well with the measured irradiance. A match was considered to be acceptable when the differences between the compared spectra, lay within the bound specified by the involved uncertainties. The applied exploitation technique of spectral direct UV irradiance measurements allowed obtaining a bound within which the retrieved parameter (either the ozone column or an aerosol property) is expected to lie with a relatively high probability.

As shown in section 7, the new Monte Carlo-based exploiting approach allowed not only retrieving the atmospheric parameters, but also evaluating their uncertainty. The proposed method was tested by retrieving the ozone column and the aerosol properties (the AOD and both Angström parameters) from direct UV irradiances measured by using the fully characterized double monochromator-based IMUK spectroradiometer.

3. New Monte Carlo-based Method

In general, Monte Carlo-based methods for uncertainty evaluations (see ISO, 2004) imply the recursive simulation of a *measurement model* that relates the inputs to the outputs. Although a *measurement model* can be as simple as a single equation, it can also be something more complex like a set of successive mathematical operations involving several measurements. In the case of the UV index calculation for example, not only the raw experimental irradiances are inputs; some calibrations factors (which involve independent measurements) are additional inputs needed to compute reliable UV indexes.

The recursive simulation of the *measurement model* requires randomly generating a large set of input values; each value rendered by a measurement model –run during a Monte Carlo simulation- must stand for a *possible* value of the output quantity. In order to ensure this, the input values must be *properly* generated; if the inputs are generated according to the available information (i.e. within their corresponding uncertainty bounds), the histogram of the outputs rendered by the model allows inferring the dispersion of possible outputs. The latter can in turn be used to evaluate the uncertainty (taken as the standard deviation of the output data).

Despite its simplicity, this Monte Carlo-based technique enables fully describing the uncertainty propagation through any *model* and comprehensively accounting for the influence of quantities that may be nonlinearly linked

3.1. Uncertainty propagation[3]

The uncertainty is a parameter that characterizes the dispersion of values that can reasonably be attributed to a measurand. Operationally, the dispersion of values of some quantity Q is described by a probability density function (PDF), $f(q)$. If the

---

[3] Section 3.1 was adapted from Cordero et al, 2008a

PDF of Q is available, its estimate and its associated standard uncertainty are taken as being equal to the expected value and the standard deviation of the PDF, respectively (Cordero and Roth, 2005).

Although obtaining the most appropriate PDF for a particular application is not straightforward, if the measurand Q is related to a set of other quantities $\mathbf{P} = (P_1,...,P_{n_P})^T$ through a *measurement model* $Q = M(\mathbf{P})$, the standard uncertainty of Q can be expressed in terms of the standard uncertainties of the *input quantities* $(P_1,...,P_{n_P})$ by using a Monte Carlo-based uncertainty propagation schema (see figure 3.1).

Note that in the proposed schema depicted in figure 3.1, the measurement model, compactly written as $Q = M(\mathbf{P})$, does not necessarily stand for single equation that renders the value of Q; the model can stand for a set of successive mathematical operations or even for a complex procedure that allows computing Q by using the input quantities in vector $\mathbf{P} = (P_1,...,P_{n_P})^T$.

The proposed Monte Carlo-based technique requires first assigning Probability Density Functions (PDFs) to each input quantity in vector $\mathbf{P} = (P_1,...,P_{n_P})^T$. Next, a computer algorithm is set up to generate an input vector $\mathbf{p_1} = (p_{1,1},...,p_{1,n_P})^T$; each element $p_{1,j}$ ($j=1,2,...,\ n_P$) in this vector is generated according to the PDF that describes the corresponding quantity P_j. By applying the generated vector $\mathbf{p_1}$ to the model $Q = M(\mathbf{P})$, the corresponding output value q_1 can be computed. If the simulating process is repeated N times ($N \gg 1$), the outcome is a series of indications $(q_1,...,q_N)$ whose *frequency* distribution allows identifying the PDF of Q. Then, irrespective of the form of this PDF, the estimate q_e and its associated standard uncertainty $u(q_e)$ can be calculated as

$$q_e = \frac{1}{N} \sum_{l=1}^{N} q_l \,,$$

(3.1)

and

$$u(q_e) = \left(\frac{1}{(N-1)} \sum_{l=1}^{N} (q_l - q_e)^2 \right)^{1/2} .$$

(3.2)

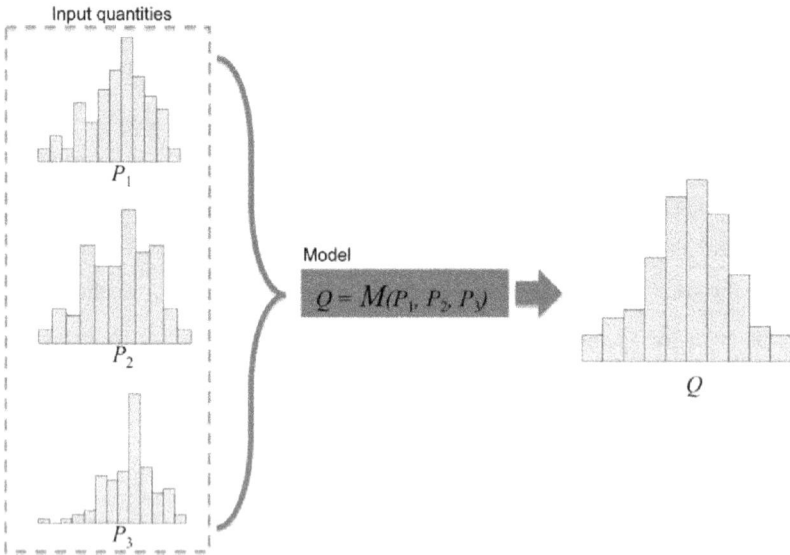

Figure 3.1.: Proposed Monte Carlo-based uncertainty propagation schema. This method allows efficiently propagating the uncertainty through any model. It implies first generating randomly a large set of values of the input quantities. Next, it requires sequentially evaluating the output quantity (which is determined by the input quantities through a measurement model). Finally, the dispersion of the computed output values can be used to evaluate both the estimate and the uncertainty of output quantity. The latter is taken as being equal to the standard deviation of the set of computed data.

The input quantities $(P_1,...,P_{n_p})$ in model $Q = M(\mathbf{P})$ are normally primaries. This means that their corresponding PDFs should be inferred by measuring directly and repeatedly the value of P_j. This well known frequency-based approach is referred to as the conventional or 'type A' analysis (Lira, 2002). However, the repeatability conditions cannot be achieved for meteorological and climatological parameters because of the temporal variability of these quantities. In this case, a situation of *information shortage* arises and information other than experimental data should be considered to assign the PDF to P_j and to evaluate its uncertainty.

In the context of information shortage, there is an internationally accepted criterion for assigning a PDF to the value of a primary quantity; this is referred to as Principle of Maximum Entropy (PME) (Cordero and Roth, 2004) and it consists of selecting the one that is most probable, among all possible PDFs that comply with the restrictions imposed by the available information. For example, if the estimate p_j and the standard uncertainty $u(p_j)$ of a quantity P_j is available, the recommended PDF for P_j is a Gaussian centred at p_j and standard deviation equal to $u(p_j)$. Instead, if only an error bound d_j can be associated with an available value p_j of the quantity P_j, the recommended PDF for p_j is rectangular over the interval $(p_j - d_j, p_j + d_j)$; then, according to (Lira, 2002), the standard uncertainty associated with p_j is

$$u(p_j) = \frac{d_j}{\sqrt{3}}. \tag{3.3}$$

This Monte Carlo-based technique was used to evaluate the uncertainties associated with the UV spectra, computed by using an 1-D radiative transfer model (see section 4), and measured by using computed by a state-of-the-art spectroradiometer (see section 5).

3.2. Integral uncertainty evaluation[4]

The Monte Carlo-based technique described in section 3.1 can used to describe the uncertainty propagation when computing experimental integrals.

Consider a two-dimensional set of J points (x_j, y_j) (where $j=1,2...,J$) formed with the elements of the vectors $\mathbf{x}=(x_1, \cdots, x_J)^T$, and $\mathbf{y}=(y_1, \cdots, y_J)^T$. The numeric integration of this set of points involves first constructing an Interpolating Function, $y=g(x)$; then, the Interpolating Function is integrated into a given range to obtain the result. The latter is referred to as an *experimental* integral if the set of J involved points are established through measurements.

The integral Q depends on the interpolating function and in turn on the set of J points (x_j, y_j). If the integral corresponds to a measurand, according to the generalized formalism discussed above, the integral calculation becomes a *measurement model* that allows evaluating the *output quantity* (the integral Q) from the values of the *input quantities*: $\mathbf{P}=(x_1, \cdots, x_J, y_1, \cdots, y_J)^T$. Hence, there are $n_P = 2J$ input quantities and a single output quantity: the integral.

Note that in this case the measurement model, compactly written as $Q = M(\mathbf{P})$, stands for a set of successive mathematical operations rather than a single equation; this set of operations involves first constructing an Interpolating Function to approximate the underlying function that produced the data, and then, integrating the Interpolating Function to obtain the result.

It is argue that the integral uncertainty can be evaluated by applying the Monte Carlo-based uncertainty propagation technique described above. This technique requires

---

[4] Section 3.2 was adapted from Cordero et al, 2008b

first assigning Probability Density Functions (PDFs) to each input quantity $(x_1, \cdots, x_J, y_1, \cdots, y_J)$ and then using the assigned PDFs to generate N independent sets of data: $[(x_{1,1}, \ldots, x_{J,1}, y_{1,1}, \ldots, y_{J,1}) \cdots, (x_{1,N}, \ldots, x_{J,N}, y_{1,N}, \ldots, y_{J,N})]$, where $N \gg 1$. These sets of data can be sequentially interpolated by using N functions and integrated into a given range to obtain a set of N results: $(q_1, \ldots, q_N)$. The latter can be in turn used to calculate the estimate and the standard uncertainty of the integral by applying equations (3.1) and (3.2), respectively. The generalization of this procedure to a three-dimensional set of points is straightforward.

The standard uncertainty rendered by equation (3.2) is reliable if the model $Q = M(\mathbf{P})$ is accurate. The calculated integral Q is accurate if the J integrated points (x_j, y_j) carry enough information about the underlying function that produced the data. As discussed below, the lack of significant biases in the integrals calculated from spectrally resolved measurements of irradiance is ensured if both the spectral resolution and the spectral period are adequately small. Note that the value of J depends on the spectral period of measurement. This means that, the length of the vector $\mathbf{P}=(x_1, \cdots, x_J, y_1, \cdots, y_J)^{\mathrm{T}}$, associated with the spectral period of measurement, determines the quality of the model outcomes $Q = M(\mathbf{P})$ and in turn the consistency of the standard uncertainty rendered by equation (3.2). The degree of the used Interpolating Function, being selected either arbitrarily or in accordance with some physical, numerical or statistical criteria, has minor importance if the value of J is adequately great. Instead, an inadequate spectral resolution can lead to systematic errors in the model output. In a context of significant biases, additional input quantities (corrections) should be introduced in the measurement model. In this situation, the uncertainty of the new correcting quantities should be also considered in the uncertainty propagation leading in turn to a greater output uncertainty.

Although, the Monte Carlo-based technique described in this section can be applied to any experimental integral, as an example, it was used to evaluate the uncertainties of UV indexes computed by integrating quality-ensured spectral UV irradiances (see section 6).

4. Uncertainty Analysis of Radiative Transfer Models[5]

As a first example, the proposed Monte Carlo-based technique was applied to the UVSPEC radiative transfer model that was taken to be a *measurement model*: radiative transfer models allow calculating the spectral UV irradiance from a set of measured *input quantities* linked with the surface reflectivity, the solar zenith angle, the ozone column and the characteristics of clouds and aerosols; the spectral irradiance yielded by a model is influenced by errors in the measurement of these input quantities.

The recursive simulation of this measurement model allowed computing the uncertainty of surface UV irradiances rendered by the model under different conditions. The uncertainty was found to be significantly dependent on the pollution; if quality-ensured inputs are available to feed the model, the expanded UV-B uncertainties (at 300 nm) under cloudless conditions can be up to 18% for clean sites, and up to 40% for sites with very large aerosol load.

4.1. Radiative Transfer Models

These models allow solving by numerical means the equation of radiative transfer that governs the transfer of radiant energy in the atmosphere. The spectrally resolved UV solar irradiance rendered by these models depends on the radiative properties (absorption, emission, scattering) of the gaseous and particulate matter of the earth's atmosphere.

The UV absorption in the atmosphere is mostly due to the oxygen and ozone molecules. Ozone strongly absorbs in the UV-B spectrum. In order to characterize

---

[5] Section 4 was adapted from Cordero et al, 2007[a], and from Cordero et al, 2007b

the gases in the atmosphere, the radiative transfer models can be loaded with several standard profiles (Anderson et al, 1986). These profiles include information of the pressure, temperature and density of the gases at different layers of the atmosphere. Normally, the models allow scaling these profiles by changing the total column value of the trace gas.

The effect of aerosols on UV irradiance is complex, owing to the variety of aerosol composition in the atmosphere, as well as its varying distribution. Because of both scattering and absorption, aerosols lead to an attenuation of the radiant energy that is expressed by the extinction coefficient $\sigma_e = \sigma_a + \sigma_s$, defined by the sum of the absorption and the scattering coefficients. The importance relative of absorption in the attenuation due to aerosols is expressed through the single scattering albedo $\omega = \sigma_s / \sigma_e$.

The extinction coefficient, the single scattering albedo, and the phase function (which depends on the scattering angle) characterize the aerosols effect. The phase function is generally obtained by using the Mie theory, which relies on classical electromagnetic equations with continuity conditions at the boundary between the particle and its surroundings. This theory leads to complex phase functions that are commonly taken as approximately equal to the Henyey-Greenstein function. The latter depends on a single parameter: the asymmetry factor g (Lenoble, 1993).

If the values of σ_e, ω and g are known at each layer and at each wavelength, the radiative transfer problem with aerosols becomes completely determined. However this is an unusual situation. Normally, because of the scarcity of experimental information, the values of ω and g are set, for all wavelengths and altitudes, to standard constant values that depend on the aerosol type.

Although for rough estimations, the value of σ_e (which depends on the aerosol concentration) can also be set to a constant, there are available several standard aerosols profiles (Shettle, 1989; Dubovik et al, 2000). Under different environments (urban, industrial, maritime, etc), these profiles provide the optical depth (*AOD*) of different layers Δy of the atmosphere. The optical depth is linked with the extinction coefficient σ_e at each layer through the expression $AOD = \sigma_e \Delta y$. The spectral influence can be also included in the computation by using the Angström's law: $AOD = \beta \lambda^{-\alpha}$, where λ is the wavelength in μm, and α and β are referred to as Angström parameters, that can be experimentally determined from a set of values of *AOD*, measured at different wavelengths (Holben et al, 1998).

4.2 Uncertainty Evaluation

4.2.1 Measurement Model

As an example, it was evaluated the uncertainty of the spectrally resolved solar UV irradiance rendered by a radiative transfer model. These computational models allow solving by numerical means the equation that governs the transfer of radiant energy in the atmosphere.

The selected model was the libRadtran software package; this is a set of programs for radiative transfer calculations whose main tool is the UVSPEC model (Mayer and Kylling, 2005). It was selected as radiative transfer solver the pseudospherical version of the DISORT solver as described in (Dahlback and Stamnes, 1991). The selected solver is refereed to as SDISORT in the libRadtran software package and is based on the one-dimensional radiative transfer solver by (Stamnes et al, 1988).

The UVSPEC model allows calculating the output quantity, that is the spectral irradiance I in the range 280-400 nm (i.e. UV irradiance), from a set of measured

33

input quantities linked with the concentration of atmospheric constituents, the surface reflectivity as well as the spectral characteristics of cloud and aerosol modulation. Under cloudless sky conditions, the quality of model outcomes has been checked by a systematic long-term comparison of spectral UV measurements and modelling results (Mayer et al, 1997).

Although efforts have been reported on the uncertainty estimation of the spectral irradiance rendered by models (Weihs and Webb, 1997; Schwander et al, 1997), they were mainly focused on the evaluation of the uncertainty associated with some typical input quantities (ozone column, optical depth, albedo, etc). The uncertainty propagation techniques applied in these works were unable to account for the nonlinear effects on the irradiance due to some uncertainty sources. This fundamental drawback may lead to overestimate the uncertainty.

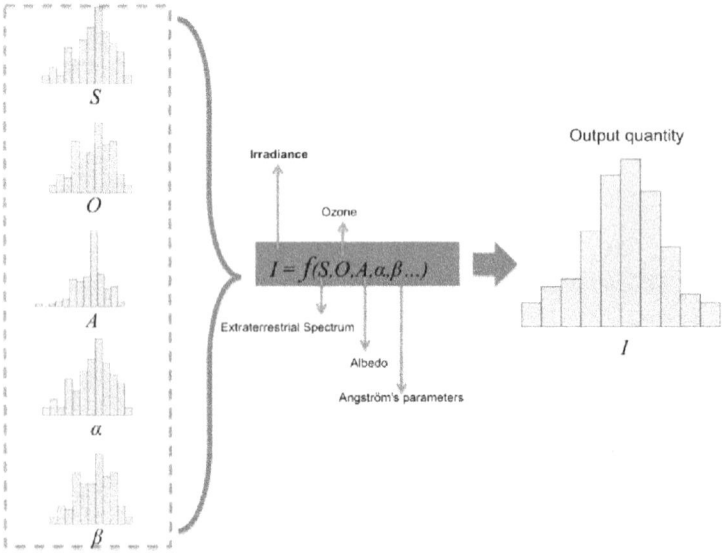

Figure 4.1.: Monte Carlo-base uncertainty propagation schema.

4.2.2. Uncertainty Sources

Table 4.1 shows the estimates p_j (where $j=1,...,8$) and the corresponding error bounds d_j, assigned to the input quantities needed to run the model: $(S_o,A,\theta,O,\omega,g,\alpha,\beta)$. Cloudless sky conditions at a sea level location were assumed and therefore, no parameter standing for the cloud characteristics was considered in the set of input quantities. The extraterrestrial spectrum S_o was quoted in the range 280-400 nm from (Gueymard, 2004). Moreover, the UVSPEC package enabled correcting the radiation quantities for the Sun-Earth distance by specifying the Julian day; the Julian day to 160 that corresponds to June 9th was set. Although the solar zenith angle θ can efficiently calculated at any geographical location by specifying the local time, the other input quantities: the albedo A, ozone column O, and the parameters used to stand for the aerosol influence (the single scattering albedo ω, the asymmetry factor g, and the Angström parameters α and β) can be estimated from ground-based measurements.

In order to be run, the UVSPEC model requires also information on the atmospheric constituent profiles. The atmospheric profile, described by Anderson et al, (1986) was selected. This profile was scaled by setting the value of the ozone column. Although the program allows also setting the density profiles of various traces gases (including ozone), if the ozone column value is maintained constant, the effect on the spectral irradiance (280-400 nm) of changing the selected profile was found to be small.

As aerosol model, that shown in (Shettle, 1989) was set. This is the default aerosol model of the libRadtran package and it loaded the properties corresponding to a rural type aerosol in the boundary layer, background aerosol above 2 km and spring-summer conditions. The selected aerosol model was scaled by setting for all wavelengths and altitudes the values of ω and g indicated in table 4.1. Moreover, although the spectrally resolved profile of the extinction coefficient was not available, by setting the Angström parameters α and β to the values shown in table 4.1, the spectral change of the extinction coefficient was included in the computation.

Table 4.1[6]. Estimates and corresponding error bounds of the input quantities utilized to run the UVSPEC model.

j	Input quantity	P_j	Estimated value p_j	Error bound d_j
1	Extraterrestrial Spectrum	S_o	Gueymard 2004*	5%
2	Albedo	A	0.2	25%
3	Solar zenith angle	θ	35°	0.2°
4	Ozone column	O	304 DU	5%
5	Single scattering albedo	ω	0.8	0.05
6	Asymmetry factor	g	0.6	0.05
7	Angström's	α	1.6	0.04
8	parameters	β	0.5	0.04

* Gueymard, C.A. The sun's total and spectral irradiance for solar energy applications and solar radiation models. Solar Energy 76, 423–453, 2004

The selected estimates of the input quantities ω, g, α, and β, (se table 4.1) correspond to a polluted scenario; in order to ensure their consistency, these aerosol properties were quote from (Dubovik et al, 2000a), which reports on aerosol data observed by worldwide contributors of the Aerosol Robotic Network AERONET (Holben et al, 1998), a well known network of remote sensing aerosol. Because the selected aerosol model was strongly scaled by specifying values for ω, g, α and β,

---

[6] Table 4.1 was quoted from Cordero et al, 2007b

the effect on the irradiance values of changing the selected aerosol optical depth profile was found to be small.

Note that the estimates shown in table 4.1 stand for a particular scenario; these values cannot be considered to be *standard estimates* because of the inherent temporal variability of the meteorological and climatological parameters.

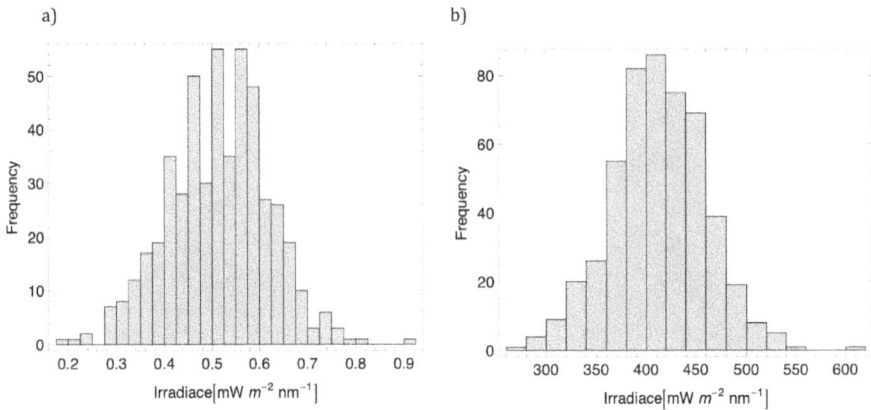

Figure 4.2[7]. Dispersion of computed values of the global irradiance at (*a*) λ=300 nm, and at (*b*) λ=400 nm. The expected values and standard deviations of these set of values are (*a*) 0.51 mW/m^{2}nm and 0.10 mW/m^{2}nm, (*b*) 414.44 mW/m^{2}nm and 45.88 mW/m^{2}nm, respectively.

---

[7] Figure 4.2 was adapted from Cordero et al, 2007b

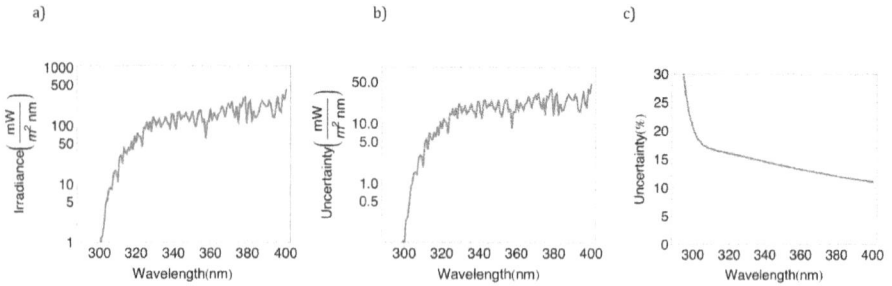

Figure 4.3[8]. (*a*) Best estimate of the global irradiance rendered by the UVSPEC model with the input data indicated in table 4.1. (*b*) Standard uncertainties of the global irradiances shown in (*a*). (*c*) Relative standard uncertainties of the global irradiances calculated with the data shown in (*a*) and (*b*).

4.2.3 Input PDFs

A Monte Carlo-based technique of uncertainty propagation (see figure 4.1) requires assigning PDFs to each input quantity P_j. Table 4.1 shows the estimate p_j and the corresponding error bound d_j of the input quantities used to evaluate the irradiance I. In each case, the value of d_j was taken to be equal to the maximum reasonable error for the available datum of the input quantity.

If the input quantities $(S_o, A, \theta, O, \omega, g, \alpha, \beta)$ are measured, the error bounds d_j associated with the estimates p_j, should not be estimated considering the temporal variability of these input quantities. The error bounds and sequentially the standard uncertainties $u(p_j)$, depend on the measurement conditions; instead, the variability depends on the measurand. If the measuring instrument is changed, the uncertainty can change even in the case of a stable quantity; on the other hand, the measurements of

---

[8] Figure 4.3 was adapted from Cordero et al, 2007b

38

meteorological or climatological quantities can change with the time but, if the measurement conditions are invariable, the uncertainty associated with these measurements can remain constant. Therefore, the values of d_j in table 4.1 stand for the maximum reasonable errors considering the measurement conditions described below.

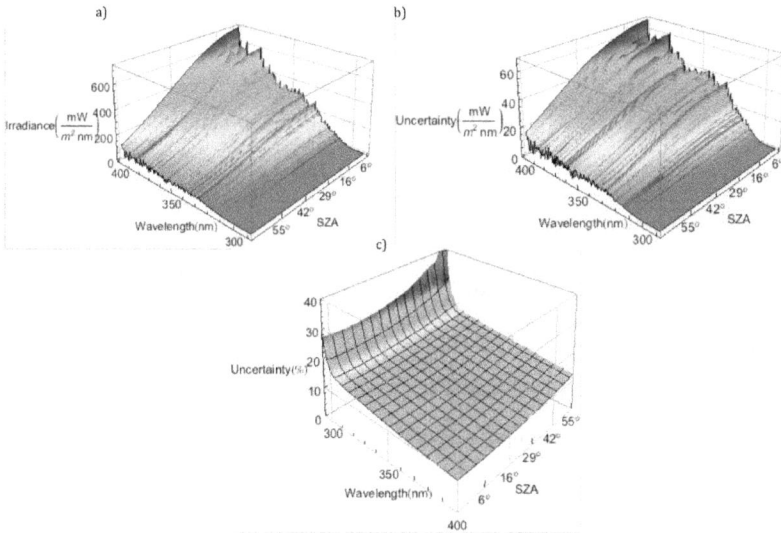

Figure 4.4[9]. (*a*) Global irradiance rendered by the UVSPEC model with the input data indicated in table 4.1 but with different values of the solar zenith angle θ; the estimates and error bounds of the other input quantities were maintained as indicated in table 4.1. (*b*) Standard uncertainties of the global irradiances shown in (*a*). (*c*) Relative standard uncertainty of the global irradiances calculated with the data shown in (*a*) and (*b*).

---

[9] Figure 4.4 was adapted from Cordero et al, 2007b

The error bound associated with each input quantity was estimated according to the characteristic and limitations of its eventual measurement. The measurement of the solar zenith angle θ and the ozone column O (by using a Microtops II instrument) are relatively simple and they could be performed with a relatively small error. However, the estimation of the other input quantities is more difficult and it requires performing some approximations.

Although the UVSPEC program allowed setting the spectrally resolved albedo $A(\lambda)$, such a detailed information is not always available. Therefore, the value of A was set to 0.2, for all wavelengths. The albedo is often determined by using a broadband instrument; accordingly, an error bound equal to 25% seems reasonable.

As indicated above, the values of ω, g, α and β were assumed to be retrieved from the data rendered by an AERONET contributor. The quality of aerosols optical properties retrieved from this Network has been analyzed by extensive sensitivity simulations (Dubovik et al, 2000b), studying the effects of both random measurement errors and systematic instrumental offsets for several aerosol models. The error bounds of ω, g, α and β were assigned considering the values reported in (Dubovik et al, 2000b) but applying a security factor such that, the error bounds in table 4.1 define intervals within which these parameters are expected to lie with a relatively high probability.

Although several extraterrestrial reference spectra have been recently reported (Gueymard, 2004; Thuillier et al, 2003; Gueymard et al, 2002), the comparisons between these spectra carried out by (Gueymard et al, 2006) have shown that the process of developing a reference spectra is reaching a maturation point; only relatively minor changes in the spectrum are expected in the future. The agreement in the spectra compared by (Gueymard et al, 2006) allowed estimating a maximum *systematic* error equal to 5% in the selected spectrum. A systematic error affects the whole set of irradiance values in S_o such that the spectrum can be slightly biased.

For each input quantity, its corresponding error bound defines an interval $(p_j - d_j, p_j + d_j)$ that should contain the value of the measurand. Because there is not additional information on the values of each quantitiy P_j than that shown in table 4.1, attending to the principle of maximum entropy (PME) (Cordero and Roth, 2004), a rectangular PDF over the interval $(p_j - d_j, p_j + d_j)$ was assigned to each input quantity; this means that the corresponding standard uncertainty associated with each input value p_j can be calculated by equation (3.3).

4.2.4 Computer Simulation

The UVSPEC program can be considered a *measurement model* $Q = M(\mathbf{P})$ that allows calculating the output quantity $Q = I$ from the values of other quantities: $\mathbf{P} = (S_o, A, \theta, O, \omega, g, \alpha, \beta)^T$. As indicated in section 3, because this measurement model $I = M(\mathbf{P})$ is strongly nonlinear, the standard uncertainty of the output quantity should be expressed in terms of the standard uncertainties of the input quantities $(S_o, A, \theta, O, \omega, g, \alpha, \beta)$ by using a Monte Carlo-based uncertainty propagation technique.

The uncertainty propagation technique required setting up a computer algorithm to generate single values of the input quantities $\mathbf{P}_1 = (S_{o,1}, A_1, \theta_1, O_1, \omega_1, g_1, \alpha_1, \beta_1)^T$. Each value was generated according to the corresponding PDF assigned in section 4.2.3 Note that in this case $S_{o,1}$ stands for an extraterrestrial spectrum randomly biased up to 5% with respect to the selected reference spectrum. With the generated values of the input quantities, the global irradiance I in the range 280-400 nm was evaluated by using the UVSPEC model: $I_1 = M(\mathbf{P}_1)$. Since this simulating process and the corresponding irradiance evaluation were repeated $N = 500$ times, the series

41

$(I_1,...,I_N)$ was formed with the outcomes. Notice that each element in these sequences stands for the spectrally resolved irradiances in the indicated range. As an example, figure 4.2 shows the nearly Gaussian dispersion of the N computed values of the irradiance at two specific wavelengths: λ=300 nm *(a)*, and λ=400 nm *(b)*.

At each wavelength λ in the range 280-400 nm, the mean and standard deviations of the series $(I_1,...,I_N)$ were calculated by applying equations (3.1) and (3.2), respectively. Then, the mean was considered to be equal to the best estimate of the surface global irradiance and the standard deviation was taken as being equal to the corresponding standard uncertainty.

Figures 4.3a shows the best estimates of the spectrally resolved global irradiance I, in the range 280-400 nm; figure 4.3b shows the corresponding standard uncertainties of the global irradiance $u(I)$. Figure 4.3c shows the relative standard uncertainty of the global irradiance, $u(I)/I$; the latter plot was built up using the data shown in figures 4.3a and 4.3b. Although in the UV-B part of the spectrum ($\leq$315 nm), lower uncertainty values were computed, the greatest relative uncertainty was calculated in this zone.

Note that the *standard* uncertainty defines a bound within which the irradiance is expected to lie with a certain probability; because of the nearly Gaussian dispersions shown in figure 4.2, if the half-width of the bound is taken to be equal to the *standard* uncertainty, the irradiance should be in this interval with a probability of about 68%. The *expanded* uncertainty of the irradiance, $U(I)$, can be calculated from the standard uncertainty $u(I)$ by applying a coverage factor equal to 2, such that $U(I)=2u(I)$. This coverage factor defines a bound within which the irradiance is expected to lie with a probability equal to about 95%.

4.3. Main influences

The solar zenith angle θ is the most important factor in determining surface irradiance, because it determines the path length through the atmospheric ozone and other absorbers and scatterers and it varies during the day and throughout the year more than any atmospheric constituent. Although it was observed that the value of θ affects also the values of $u(I)$, the change induced on the relative uncertainty $u(I)/I$ was found to be comparatively small. Figures 4.4a shows the best estimates of the spectrally resolved global irradiance I, in the range 280-400 nm; this map was built up calculating the spectral irradiances with different values of θ; the error bound of θ was not changed and it was taken every time as being equal to 0.2° (see table 4.1); the estimates and error bounds of the other input quantities were maintained as indicated in table 4.1. Figure 4.4b shows the standard uncertainties $u(I)$ of the global irradiances depicted in figure 4.4a. Figure 4.4c shows the relative standard uncertainty of the global irradiance, $u(I)/I$; the latter plot was built up using the data shown in figures 4.4a and 4.4b.

Instead of the solar zenith angle, the estimate of the Angström parameter α affected considerably the values of $u(I)/I$. Figure 4.5 show the relative standard uncertainty of I, calculated with different estimates of the Angström's parameter α. In these calculations, the error bound of α was not changed and it was taken every time as being equal to 0.04 (see table 4.1); the estimates and error bounds of the other input quantities were maintained as indicated in table 4.1. It can be observed in figure 4.5 that, despite the wavelength, $u(I)/I$ diminished with the value of α. Note that greater values of α characterize polluted air. It is concluded that the relative uncertainty of the irradiance strongly depends on the aerosol conditions.

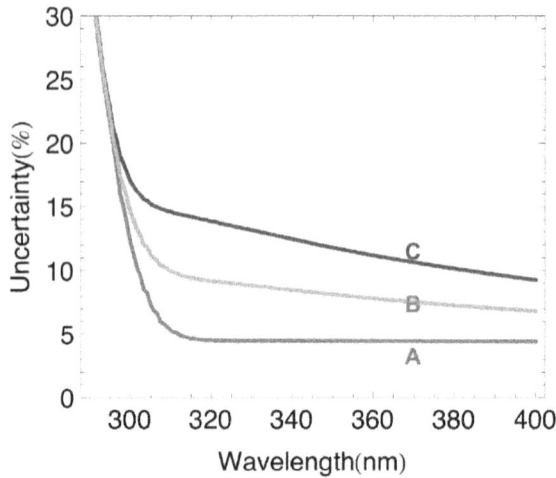

Figure 4.5[10]. Relative standard uncertainties of the global irradiances rendered by the UVSPEC model considering that the estimates of the Angström's parameter α was A: 0.6; B: 1.1 and C: 1.6; the other estimates of the input quantities and all the error bounds were taken as indicated in Table 4.1.

It is apparent from figure 4.5 that the relative uncertainties can significantly change with estimates of the involved input quantities. This means that the uncertainty results are restricted such that they can be considered valid only for the conditions detailed in table 4.1. This limitation cannot be overcome because of the great differences between the input values that correspond to different meteorological conditions.

---

[10] Figure 4.5 was adapted from Cordero et al, 2007b

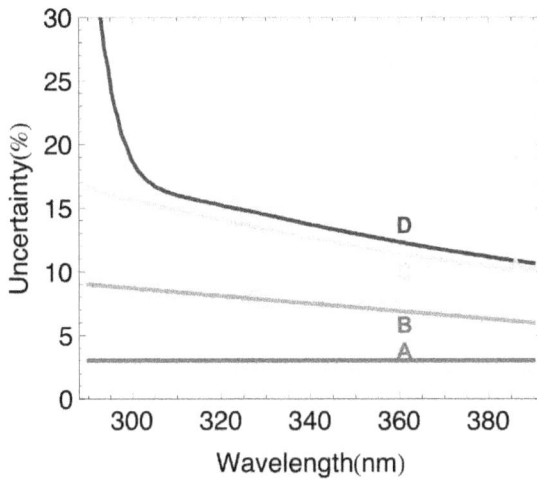

Figure 4.6[11]. Contributions to the relative standard uncertainty of the global irradiance I calculated by using the input information shown in table 1; the standard uncertainties were calculated considering A: only the uncertainty associated with the extraterrestrial spectrum S_o; B: the uncertainties of both S_o and the Angström parameter β; C: the uncertainties of S_o, β and the single scattering albedo ω; D: the uncertainties of S_o, ω, β, α and the ozone column O.

Figure 4.6 shows the contributions to the relative standard uncertainty $u(I)/I$ shown in figure 4.3c. The values of $u(I)/I$ in this plot were calculated considering A: only the uncertainty associated with the extraterrestrial spectrum S_o; B: the uncertainties of both S_o and the Angström parameter β; C: the uncertainties of S_o, β and the single scattering albedo ω; D: the uncertainties of S_o, ω, β, α and the ozone column O. It is apparent from figure 4.6 that the uncertainties of the 4 parameters (S_o, β, ω, O) accounted for practically the total uncertainty $u(I)/I$, under the conditions characterized by the data shown in table 4.1; in the UV-A part of the spectrum (315-

---

[11] Figure 4.6 was adapted from Cordero et al, 2007b

45

400 nm), the main contributors to $u(I)/I$ were the uncertainties attributed to the single scattering albedo ω and the Angström parameter β. Instead, in the UV-B part of the spectrum (280-315 nm), the irradiance uncertainty was also strongly dependent on uncertainty associated with the ozone column datum.

5. Uncertainty Analysis of double monochromator-based spectroradiometers[12]

The reference instruments to measure the surface UV irradiance are based on double monochromator systems. The spectral irradiances yielded by these instruments are affected by temporal instabilities and nonlinearities in the signal, as well as uncertainties introduced in the needed prior calibrations.

By using the Monte Carlo-based technique described in section 3, below it has been carried out an uncertainty analysis of the spectral irradiances measured by using the spectroradiometer of the Leibniz Universität Hannover (Institut für Meterologie und Klimatologie, IMUK). This instrument complies with the requirements of the Network for the Detection of Atmospheric Composition Change (NDACC). The spectral measurements were performed under cloudless sky conditions at the Izaña observatory (28.3° N, 16.5° E, 2367 m above sea level, Tenerife, Spain), during an international intercomparison campaign organized in the framework of the project Quality Assurance of Spectral Ultraviolet Measurements in Europe (QASUME).

It was found that despite the variations due to wavelength shifts, the relative *expanded* uncertainty was about 6% in the UV-A part of the spectrum; an increment was observed at wavelengths shorter than 315 nm such that the *expanded* uncertainty of the UV-B irradiance at 300 nm wavelength was about 9%. It was also found that the uncertainties involved in the absolute calibration procedure accounted for about 65% of the UV-A uncertainty. Although only a double monochromator was analyzed, the methodology applied to evaluate the uncertainty is general and agrees with recommendations of the ISO *Guide to the Expression of Uncertainty in Measurement*.

---

[12] Section 5 was adapted from Cordero et al, 2008a

5.1. Spectroradiometer systems

The spectrally resolved irradiance can be efficiently measured by using spectroradiometer systems. All these instruments render the results of a solar scan as a two dimensional set of J points (λ_j, S_j), where $(j=1,2,...,J)$; S_j is the measured value of the signal registered at the wavelength λ_j. The value of J is determined by the spectral period of the measurements (i.e. the smallest difference between the wavelengths corresponding to two adjacent measurements in the set of data).

The values of λ_j and S_j can be affected by several uncertainty sources; if these errors are corrected, a new set of J points (λ'_j, S'_j) can be formed. The irradiances E_j can be then computed from the *corrected* signals S'_j as

$$E_j = \left(\frac{S'_j}{r_j} \right),$$

(5.1)

where r_j is the spectral responsivity of the spectroradiometer (evaluated at the wavelength λ_j).

The responsivity r_j is determined by carrying out an absolute calibration. The calibration involves relating the primary spectral irradiance $E_{j,c}$ (obtained from the certificate of a tungsten halogen lamp used as calibration source at each wavelength λ_j) to the signal values obtained by scanning the lamp; as indicated above, the measurements are rendered as a set of J values $(S_{1,c}, \cdots, S_{J,c})$ (the subscript c refers to the calibrating lamp). The correction of the errors affecting each element $S_{j,c}$ leads to the *corrected* signal values $S'_{j,c}$ that can be then used to evaluate the responsivity as

$$r_j = \left(\frac{S'_{j,c}}{E_{j,c}} \right). \qquad\qquad (5.2)$$

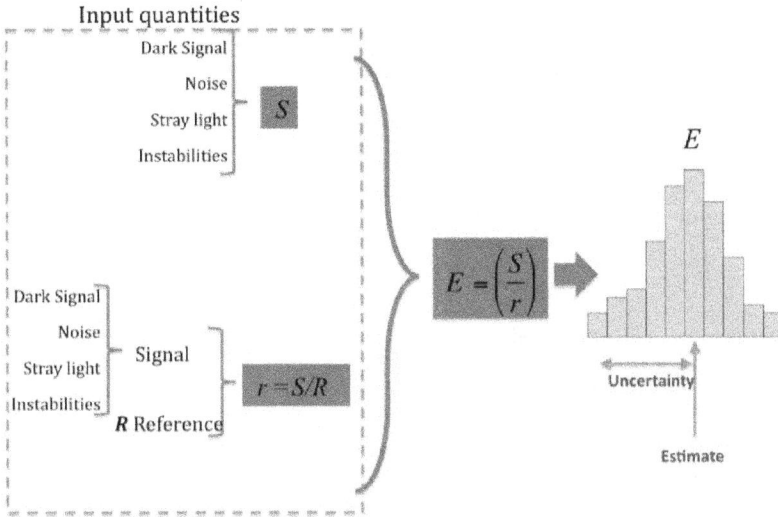

Figure 5.1.: Monte Carlo-base uncertainty propagation schema.

5.2 Uncertainty Evaluation

In what follows, by applying the Monte Carlo-based method described in section 3.1, it is computed the uncertainty associated with some spectral irradiance measurements performed at Izaña observatory (Tenerife, Spain) by using the spectroradiometer system of the Leibniz Universität Hannover (Institut für Meterologie und Klimatologie, IMUK). The measurements were performed during an international intercomparison campaign organized in the framework of the QASUME project. The double monochromator-based IMUK spectroradiometer is a NDACC-certified instrument whose characterization has been reported by (Cordero et al, 2008a).

5.2.1 Uncertainty Sources

A scan render a set of J points (λ_j, S_j), where $j=1,2,\ldots,J$. The values of the wavelength λ_j indicated by the display device of the spectroradiometer can be slightly shifted. The misalignment can be corrected by carrying out a wavelength calibration which involves comparing the wavelength indicated by the instrument with the known wavelength of some spectral lines of a low pressure mercury-argon Hg(Ar) pencil lamp.

Although before field measurements, the wavelength calibration allows discarding significant biases in the indicated values of λ_j, additional systematic shifts in the values of the output wavelengths can be induced by high environmental temperatures; these changes in the temperature cannot be fully counteracted even if the measuring instrument is operated within a weather-proof box. A post-measurement quality control (Slaper et al, 1995) has allowed detecting systematic shifts up to about ±0.05 nm during field measurements performed by using the IMUK spectroradiometer. The effect of these temperatures-induced shifts on the wavelength λ_j can be characterized by using a single additive factor:

$$\lambda'_j = \lambda_j + z. \qquad\qquad (5.3)$$

Ideally, the factor z should allow counteracting the effect of the wavelength shifts, but its accurate determination is difficult. Therefore, if the spectroradiometer is operated in a weather-proof box, the value of z can be considered to be zero: Although in that case, the estimates of λ'_j are not affected by z, the uncertainty of z is not zero, and therefore, it does affect the uncertainty of λ'_j.

On the other hand, the main uncertainty sources affecting each signal S_j rendered by a NDACC-certified instrument (as the IMUK spectroradiometer) are radiometric instabilities and offset variations (see Cede et al, 2006) whose effects on each *indicated* signal value S_j, can be characterized by using an additive factor w and a multiplicative factor v:

$$S'_j = vS_j + w,\tag{5.4}$$

The values of w and v can be estimated by conducting a stability test; it involves repeatedly measuring the irradiance from a stable light source (i.e. a tungsten halogen lamp). If the stability is reasonably good and the offset is regularly counteracted, the estimates of the v and w can be taken to be 1 and 0, respectively. Although in this case, the value of v and w do not affect the estimates of S'_j, it is apparent from equation (5.4) that the uncertainties of v and w do affect the uncertainty of S'_j.

5.2.2 Measurement model

As shown above, the spectral irradiance is determined from the set of J points (λ'_j, S'_j) calculated from the points (λ_j, S_j) yielded by a spectroradiometer. The values of λ_j and S_j can be affected by several error sources; these errors can be corrected by applying equations (5.3) and (5.4) such that a new set of J points (λ'_j, S'_j) can be formed. The latter can be interpolated to approximate the UV spectrum $E(\lambda)$.

The calculation of E_j from the corrected values of S'_j (by equation (5.1)) requires the prior evaluation of the spectral responsivity by applying equation (5.2); this in turn involves additional experimental information: the J values $S_{j,c}$ obtained by scanning

51

the lamp utilized during the absolute calibration procedure, and the corresponding J values of the irradiance $E_{j,c}$ obtained from the calibration certificate of the lamp.

Attending to the formulation introduced in section 3.1, the procedure that allows retrieving the underlying UV spectrum $E(\lambda)$ from experimental data measured by using the IMUK spectroradimeter, can be then compactly represented as

$$E(\lambda) = M(\mathbf{l_j}, \mathbf{S_{j,c}}, \mathbf{S_j}, \mathbf{E_c}),$$ (5.5)

where $\mathbf{l_j} = (\lambda_1, \lambda_2, ..., \lambda_J)^T$, $\mathbf{S_{j,c}} = (S_{1,c}, \cdots, S_{J,c})^T$, $\mathbf{S_j} = (S_1, \cdots, S_J)^T$, and $\mathbf{E_c} = (E_{1,c}, \cdots, E_{J,c})^T$. Equation (5.5) stands for a *measurement model* that allows expressing the standard uncertainty of the *output quantity* (the spectrum $E(\lambda)$) in terms of the standard uncertainties of the $4J$ *input quantities* $(\mathbf{l_j}, \mathbf{S_{j,c}}, \mathbf{S_j}, \mathbf{E_c})$. This can be carried out by using a Monte Carlo-based computer simulation.

Note that in order to evaluate $E(\lambda)$ additional information is also needed: the estimate of the factor z used to characterize the wavelength shifts (this additive factor affects the values of the elements of vector $\mathbf{l_j}$ as shown in equation (5.3); and the factors v, and w used to characterize the influence of instabilities in the signal; the elements of vectors $\mathbf{S_{j,c}}$ and $\mathbf{S_j}$ are affected by different and independent values of the factors v and w, as shown in equation (5.4). Although they were not explicitly included in equation (5.5), z, v and w can be also considered input quantities in the measurement model.

5.2.3 Input PDFs

A Monte Carlo-based technique of uncertainty propagation requires assigning probability density functions (PDFs) to each input quantity in the measurement model defined above.

The signal and wavelength values are rendered by the IMUK spectroradiometer without ambiguities; therefore, each signal value in vectors $\mathbf{S}_{j,c}$ and $\mathbf{S}_j$ and each wavelength in the vector $\mathbf{l}_j$, was considered to be described by Dirac delta functions centered at the value indicated by the instrument.

The factors v and w (that characterize the instabilities and offset variations affecting the signal values in vectors $\mathbf{S}_{j,c}$ and $\mathbf{S}_j$) were described by using a normal and a rectangular PDF, respectively. The effect of radiometric instabilities was assessed by repetitively measuring the irradiance of a standard lamp; the standard deviation of the nearly Gaussian dispersion observed in the data, was about 1% such that the factor v was considered to be described by a normal PDF centered at 1 and standard deviation 0.01. Because the offset was regularly measured, its non-detected maximum reasonable variation w_{max} was relatively small; I took $w_{max} = 3 \times 10^{-12}$ A and therefore, the factor w was described by using a rectangular PDF over the interval $(- w_{max}, + w_{max})$. Since averaging was applied, the influence of noise was ignored; PMT are not significantly affected by noise. Moreover, the factor z that characterize the shifts in the values of λ_j due to changes in the temperature, was described by using a rectangular PDF over the interval (-0.05,+0.05) nm.

According to the calibration certificate, the standard uncertainties associated with the irradiance values $E_{j,c}$ of the calibrating lamp are 1.5%. However, attending to eventual errors due to the aging and variations in the current, I took the relative standard uncertainties associated with the irradiance values of lamp as being equal to 2%. Therefore, each value $E_{j,c}$ in the vector $\mathbf{E_c}$ was described by using a normal PDF centered at the available value of $E_{j,c}$ and standard deviation $(0.02E_{j,c})$.

5.2.4 Computer simulation

By using the PDFs assigned above, values of each input quantity were generated. The elements of vector $\mathbf{S_{j,c}}$ were generated by using experimental data obtained when measuring the global spectral UV Irradiance at 13:00 h local time on June 9th, 2005 at Izaña observatory (Tenerife, Spain). The elements of vector $\mathbf{S_{j,c}}$ were later modified by using single generated values of v, and w, such that, according to equation (5.4), J values $S'_{j,c}$ were calculated. The same procedure was applied to the generated elements of vector $\mathbf{S_j}$. The $S'_{j,c}$ values and the corresponding spectral irradiances $E_{j,c}$ obtained from vector $\mathbf{E_c}$ were used to evaluate responsivity values r_j by applying equation (5.2). Sequentially, the S'_j values, the calculated values of r_j allowed calculating the irradiance values E_j by applying equation (5.1).

Since both the simulating process described above and the corresponding assessment of the UV irradiance were repeated $N=300$ times, J sequences $(E_{j,1},...,E_{j,N})$ were formed corresponding to J wavelengths λ_j. Each simulation required generating single values of the input quantities $(\mathbf{l_j}, \mathbf{S_{j,c}}, \mathbf{S_j}, \mathbf{E_c})$ as well as new values of z, w, and v, and sequentially determining each time a new interpolating function $E(\lambda)$.

The elements of vector $\boldsymbol{\lambda_j}$ were used to calculate a set of J values λ'_j by using a generated value of z (see equation (5.3)). These values and the calculated values of E_j allowed building up a set of J points (λ'_j, E_j) which was linearly interpolated.

a)   b)

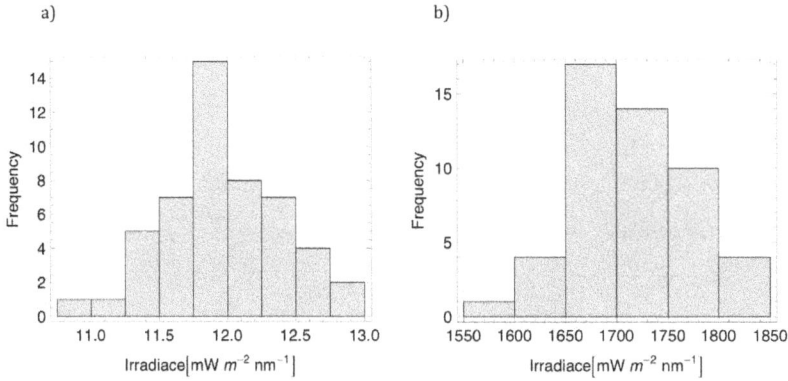

Figure 5.2[13]. Dispersion of the N possible values at 13:00 h local time. (*a*) Global Irradiance at 300 nm, (*b*) Global Irradiance at 400 nm. The expected values and standard deviations of these dispersions are (*a*) 11.9 mW/m²nm and 0.40 mW/m²nm, (*b*) 1715 mW/m²nm and 58 mW/m²nm.

Figure 5.2 shows the nearly Gaussian dispersion of the N values of the irradiances obtained by evaluating each spectrum in the series $\left(E(\lambda)_1,...,E(\lambda)_N\right)$ at two specific wavelengths of the spectrum. The standard deviations of these data can be used to assess the irradiance uncertainty. The *standard* uncertainties $u(E)$ of the irradiances at these wavelengths were taken to be equal to the standard deviations of the data shown in figure 5.2; the standard deviations were calculated by applying equation (3.2). In the same way, the standard uncertainties at each wavelength λ in the range 290-400 nm were evaluated.

Figure 5.3a shows the global spectral UV Irradiance measured at 13:00 h local time (the solar zenith angle was 5°) on June 9th, 2005 (cloudless conditions) at Izaña

---

[13] Figure 5.2 was adapted from Cordero et al, 2008a

observatory (Tenerife, Spain) by using the IMUK spectroradiometer. Figure 5.3b shows the standard uncertainties $u(E)$ of the irradiances in figure 5.3a.

Figure 5.3c shows the relative *expanded* uncertainty of the global irradiance, $U(E)/E$. The plot was built up by using the data shown in figures 5.3a and 5.3b. The *expanded* uncertainty $U(E)$ was calculated from the standard uncertainty $u(E)$ by applying a coverage factor equal to 2, such that $U(E)=2u(E)$. Note that the uncertainty values can be used to define a bound within which the irradiance is expected to lie with a certain probability. Because of the nearly Gaussian frequency distributions shown in figure 5.2, if the half-width of the bound is taken to be $2u(E)$, the irradiance should be in this interval with a probability of about 95%.

a)   b)   c)

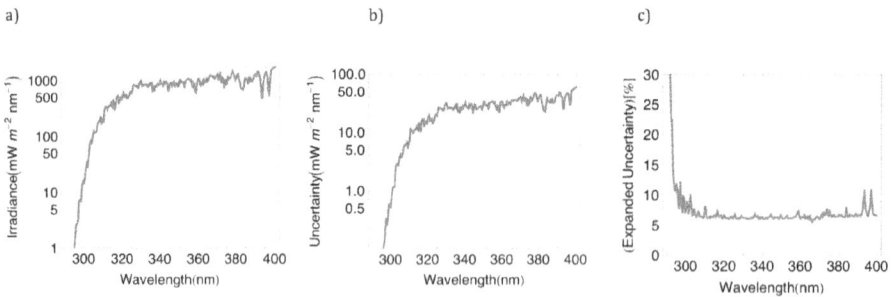

Figure 5.3[14]. *(a)* Global spectral UV Irradiance measured at 13:00 h local time (the solar zenith angle was 5°) on June 9th, 2005 (cloudless conditions) at Izaña observatory (Tenerife, Spain) by using a spectroradiometer system of the Leibniz Universität Hannover (Institut für Meterologie und Klimatologie, IMUK).

(b) Standard uncertainties of the irradiance measurements shown in *(a)*

(c) Relative expanded uncertainties of the global irradiances shown in *(a)*.

---

[14] Figure 5.3 was adapted from Cordero et al, 2008a

5.3. Main Influences

Note that the measurements reported in figure 5.3a were obtained during a campaign. At solar zenith angles smaller than 45° (where the influence of the cosine error was small), the measurements of the other 5 teams that participated in the campaign were within the bound given by the *expanded* uncertainties; the measurements of only 3 of the other 5 teams that participated in the campaign were within the bound defined by the *standard* uncertainties; these define an interval within which the irradiance is expected to lie with a probability equal to 68%.

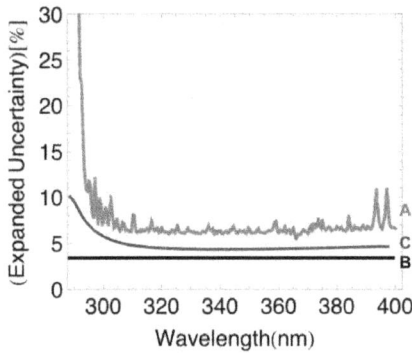

Figure 5.4[15]. Curve A depicts the relative expanded uncertainties of the irradiance values rendered by the IMUK spectroradiometer at 13:00 h ($\theta=5°$). Curves B and C show the main contributions to the overall uncertainties in curve A. The uncertainties in curve B were calculated considering only the uncertainty associated with the spectrum E_e of the calibrating lamp; instead, the uncertainties in curve C were calculated considering the uncertainties of the spectral responsivity r (which includes the uncertainty of the spectrum E_e and that corresponding to the signal values

---

[15] Figure 5.4 was adapted from Cordero et al, 2008a

obtained by scanning the calibrating lamp).

 Although at wavelengths longer than 315 nm the contributions to the uncertainty due to the temporal offset variations were small, the increment in the relative uncertainty observed in the UV-B part of the spectrum, can be attributed to the additive variations in the measured signals due to that error source. Since averaging was applied, the influence of noise was insignificant; PMT are not considerably affected by noise.

Curve A in figure 5.4 depicts the relative expanded uncertainties of the irradiance values rendered by the IMUK spectroradiometer at 13:00 h ($\theta=5°$). Curves B and C show the main contributions to the overall uncertainties in curve A. Curve B was calculated considering in the uncertainty propagation only the uncertainty associated with the spectrum E_c of the calibrating lamp; instead, curve C was calculated considering in the uncertainty propagation the uncertainty of the spectral responsivity r (which already includes the uncertainty of the spectrum E_c and that corresponding to the signal values S_c obtained by scanning the calibrating lamp). From figure 5.4, it is concluded that the uncertainties involved in the absolute calibration procedure were the main contributor to the irradiance uncertainty. At solar zenith angles smaller than 30°, where the influence of the cosine error was small, the uncertainty attributed only to the spectrum of the calibrating lamp accounted for about 60% of the uncertainty in the irradiance at 300 nm.

6. UV index Uncertainty[16]

Although the Monte Carlo-based technique described section 3 can be applied to any experimental integral, as an example, it was used to evaluate the uncertainties of the UV index. This is evaluated by calculating the integral in the range 250-400 nm of the spectral UV irradiance weighted by using the McKinlay-Diffey Erythema action spectrum. The latter describes the relative effectiveness of energy at different wavelengths in producing a particular biological response.

The spectral UV irradiances were measured by using the IMUK spectroradiometer (see section 5). The former complies with the requirements of the Network for the Detection of Atmospheric Composition Change (NDACC)). The measurements were performed during an international intercomparison campaign organized in Tenerife (Spain), in the frame of the project Quality Assurance of Spectral Ultraviolet Measurements in Europe through the Development of a Transportable Unit (QASUME).

As expected, it was found that the index uncertainty strongly depended on the uncertainties affecting spectral UV-B irradiance measurements.

6.1. UV index

The evaluation of the nondimensional UV index "I_n" requires integrating in the range 250-400 nm the spectral UV irradiance $E(\lambda)$:

$$I_n = \int_{250}^{400} W(\lambda) E(\lambda) d\lambda, \tag{6.1}$$

---

[16] Section 6 was adapted from Cordero et al, 2008b

where $A=40\left(m^2/W\right)$, and λ is the wavelength in nm. By international accord, the weighting factor is the so-called McKinlay-Diffey Erythema action spectrum (McKinlay and Diffey 1987):

$$W(\lambda)=\begin{cases} 1 & 250 \leq \lambda \leq 298 \\ 10^{0.094(298-\lambda)} & 298 < \lambda \leq 328 \\ 10^{0.015(139-\lambda)} & 328 < \lambda \leq 400 \end{cases}. \tag{6.2}$$

This function describes the relative effectiveness of energy at different wavelengths in producing a biological response.

It is clear from equation (6.1) that the UV index value is affected by the errors in the determination of the spectral UV irradiance $E(\lambda)$.

As explicated in section 5.1, the spectrally resolved irradiance can be efficiently measured by using spectroradiometer systems. All these instruments render the results of a solar scan as a two dimensional set of J points (λ_j, S_j), where $(j=1,2,\ldots,J)$; S_j is the measured value of the signal registered at the wavelength λ_j.

If the errors affecting the values of λ_j and S_j are corrected, a new set of J points (λ'_j, S'_j) can be formed. The irradiances E_j can be then computed from the *corrected* signals S'_j by using equation (5.1). As explicated in section 5.1, the responsivity r_j in turn evaluated by equation (5.2) is determined by carrying out an absolute calibration.

Afterwards, the irradiance values E_j calculated by equation (5.1) can be utilized to form a set of J points (λ', E_j), which can be used to evaluate the UV index by equation (6.1). This involves first constructing an interpolating function to approximate the underlying surface UV irradiance $E(\lambda)$ and then, integrating the interpolating function to obtain the UV index (see equation (6.1)).

6.2. Error sources in the UV index computation

The UV index value is affected by errors linked with an inadequate approximation to the *underlying* biologically weighted irradiance $E_o(\lambda)$; these errors arise if the experimental points yielded by the spectroradiometer do not carry enough information to infer satisfactorily the function $E_o(\lambda)$. This experimental information scarcity can be due to an insufficient spectral resolution and to an inadequate spectral period of measurements (i.e. the smallest difference between the wavelengths corresponding to two adjacent measurements in the set of data). These errors are not linked with the uncertainties associated with the integrated data and they remain even if the data rendered by a spectroradiometer are not uncertain.

Although the IMUK spectroradiometer can scan the solar spectrum measuring the irradiances every 0.1 nm, the spectral resolution of the equipment (as defined by Seckmeyer et al, 2001) is greater. The spectral resolution of a spectroradiometer is determined by the slit function and is taken to be equal to the Full Width of the function at a Half of its Maximum (FWHM) (Seckmeyer et al, 2001). Figure 6.1a shows the nearly Gaussian slit function of the IMUK spectroradiometer, obtained by scanning a spectral line of a Hg(Ar) pencil lamp; the FWHM of the IMUK spectroradiometer is about 0.5 nm.

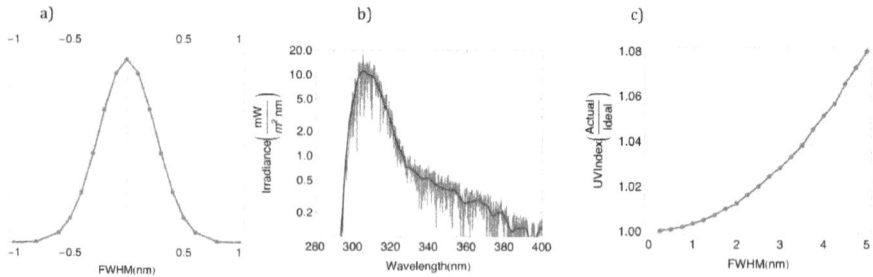

Figure 6.1[17]. (*a*) Slit function obtained by scanning with the IMUK spectroradiometer a spectral line of a pencil lamp; λ_o was 334.15 nm. (*b*) Biologically weighted UV Irradiance corresponding to the conditions observed at 10:30 h local time on June 9th, 2005 at Izaña observatory; thin line: data that it would be obtained if the irradiances were measured every 0.05 nm by using a high resolution instrument; bold line: data that it would be obtained if the irradiances were measured every 0.5 nm by using the IMUK instrument. (*c*) Ratio between the UV indexes calculated from measurements obtained (under the conditions mentioned in (*b*)) by instruments with different resolution; *ideal*: UV index that it would be obtained if the irradiances were measured every 0.5 nm by using the IMUK instrument; *actual*: UV index it would be obtained if the irradiances were measured by using a spectroradiometer with a triangular slit function and resolution FWHM (the scanning step size was also taken to be equal to the FWHM).

Most Fraunhofer lines of the solar spectrum are narrower than 0.01 nm. Therefore, the IMUK spectroradiometer render a version of the actual spectral irradiance convolved with its slit function. This effect is shown in figure 6.1b. This figure depicts the biologically weighted UV Irradiance $E_w(\lambda) = W(\lambda)E(\lambda)$ at Izaña observatory calculated by using the UVSPEC radiative transfer model (Mayer and Kylling 2005) for the conditions observed at 10:30 h on June 9th, 2005 (cloudless

---

[17] Figure 6.1 was adapted from Cordero et al, 2008b

62

conditions). The thin line in figure 6.1b stands for a set of data that it would be obtained if the irradiances were measured every 0.05 nm by using a high resolution instrument. Instead, the bold line stands for the data that it would be obtained if the irradiances were measured every 0.5 nm by using the IMUK spectroradiometer; this latter spectrum was calculated by performing a discrete convolution of the data yielded by the radiative transfer model, with the slit function of the IMUK spectroradiometer (figure 6.1a).

As shown in figure 6.1b, the slit function leads to smoothed data. This means that, although the IMUK spectroradiometer can scan the solar spectrum measuring the irradiances every 0.1 nm, the resulting adjacent measurements are correlated and the data obtained by using a spectral period smaller than the FWHM are oversampled. It has been recommended to oversample the spectrum by about half of the FWHM (Seckmeyer et al, 2001). In the case of the instruments with FWHM<1 nm, such a short spectral period of measurements will allow easily describing the underlying solar spectrum and therefore, the degree of the Interpolating Function, will not significantly affect the value of the integrals.

Although the limited spectral resolution can prevent an adequate description of the spectral variations in the underlying solar spectrum, this effect does not necessary lead to biased integral; actually, the integrals under the two curves shown in figure 6.1b, $E_w(\lambda)$ and $E_o(\lambda)$, are practically the same. Figure 6.1c shows the ratio between the UV indexes calculated for the conditions observed at 10:30 h on June 9th, 2005 at Izaña observatory. These calculations stand for the UV indexes that would be computed from the data rendered by instruments with different resolutions (see caption for details). From figure 6.1c, it is concluded that biased integrals can be calculated if irradiance measurements are performed with instruments of either low resolution or at spectral periods greater than 1 nm. Instead, a set of irradiances

63

measured by using the IMUK instrument at intervals of 0.5 nm carries enough information about the underlying surface solar spectrum, and therefore, it should allow computing reliable integrals.

6.3. Uncertainty Evaluation

Although, the proposed Monte Carlo-based technique can be applied to any experimental integral (see section 3.2), as an example, it was used to evaluate the uncertainties of the UV index. This was computed by integrating spectral measurements performed by using the IMUK spectroradiometer (see section 5).

6.3.1. Measurement models

Attending to the formulation introduced in section 3.1, the procedure that allowed evaluating the UV index from experimental data indicated above, implied first building up measurement models. In the IMUK spectroradiometer case, the corresponding model was based on that described in sections 5.2.2.

Note that in this case, the *measurement model* is not a single equation but a set of successive operations or activities involving several measurements. In addition to those operations described in section 5.5.2, supplementary activities are required in order to compute the UV index from the J points $\left(\lambda'_j, E_j\right)$: it involves first constructing an interpolating function to approximate the underlying surface UV irradiance $E(\lambda)$ and then integrating in the range 250-400 nm the biologically weighted spectral UV irradiance (see equation (6.1)).

The input quantities in the measurement model were described by using the PDFs as explained in sections 5.2.3. Note that the McKinlay-Diffey Erythema action spectrum is a definition and accordingly $W(\lambda)$ was considered to be given.

6.3.2 IMUK spectroradiometer simulation

The simulations that allowed calculating these standard uncertainties were carried out considering as input quantities the signal values S_i indicated by the measuring instrument at 13:00 h local time on June 9th, 2005, at Izaña observatory.

By using the PDFs assigned in section 5.2.3 values of each input quantity were generated. The generated elements of vector $S_{i,c}$ were modified by using single generated values of v, and w, such that, according to equation (5.4), J values $S'_{j,c}$ were calculated. The same procedure was applied to the generated elements of vector S_i. The $S'_{j,c}$ values and the corresponding spectral irradiances $E_{j,c}$ obtained from vector E_c were used to evaluate responsivity values r_j by applying equation (5.2). Sequentially, the S'_j values, the calculated values of r_j, allowed calculating the irradiance values E_j by applying equation (5.1); the weighting factor W_j was calculated by using equation (6.2).

The generated elements of vector λ_i were used to calculate a set of J values λ'_j by using a generated value of z (see equation (5.3)). These values and the calculated values of E_j allowed building up a set of J points $\left(\lambda'_j, E_j\right)$ which was linearly interpolated. The interpolating function $E(\lambda)$ was then integrated to obtain the UV index (see equation (6.1)). The value of the UV index was not significantly affected by the degree of the Interpolating Function. This was due to the adequately small spectral period of measurements.

Since both the simulating process described above and the corresponding assessment of the UV index were repeated $N=500$ times, a series $(I_{n,1},...,I_{n,N})$ was formed. Each simulation required generating single values of the input quantities $(\lambda_i, S_{i,c}, S_i, E_c)$ as well as new values of z, w, v, and sequentially determining each time a new interpolating function $E(\lambda)$.

Moreover, the *standard* uncertainty of the UV index, $u(I_n)$, was taken to be equal to the standard deviation (calculated by applying equation (3.2)) of the series $(I_{n,1},...,I_{n,N})$ Figure 6.2 shows the dispersion of the N values of the UV index in the series $(I_{n,1},..., I_{n,N})$, obtained considering as input quantities the signal values $\mathbf{s_j}$ indicated by the measuring instrument at 13:00 h local time on June 9th, 2005, at Izaña observatory (see figure 5.3a).

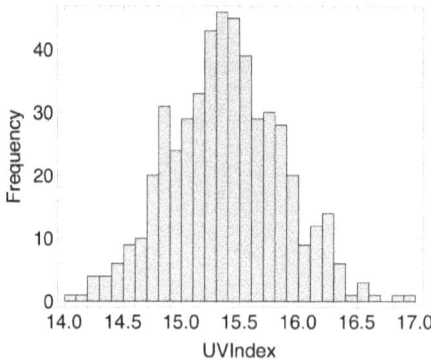

Figure 6.2[18]. Dispersion of possible values of the UV index at 13:00 h local time on June 9th, 2005 (cloudless conditions) at Izaña observatory (Tenerife, Spain) by using the IMUK spectroradiometer. The mean and the standard deviation are respectively 15.4 and 0.5.

6.4 Main influences

Because the UV index calculation implies the numeric integration of the set of J points (λ'_j, E_j), the uncertainty of both the ordinates and the abscissas should affect

---

[18] Figure 6.2 was adapted from Cordero et al, 2008b

the integral uncertainty. As expected, it was found that the uncertainty of the UV-B index strongly depended on the uncertainty sources affecting the spectral irradiance measurements (the ordinates); the influences of the uncertainty sources affecting the values of the wavelengths (the abscissas) were relatively small.

In the case of the IMUK spectroradiometer, the driving factor determining the UV-B uncertainty was the dark signal. Although at wavelengths longer than 315 nm the contributions to the uncertainty due to the temporal offset variations were small, the increment in the relative uncertainty observed in the UV-B part of the spectrum, can be attributed to the additive variations in the measured signals due to that error source.

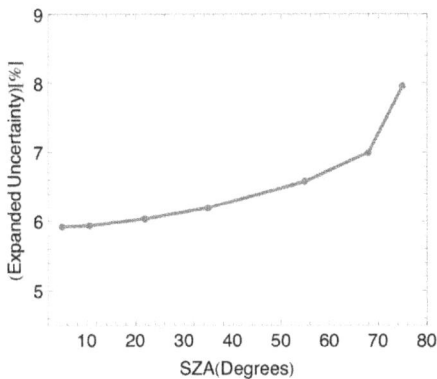

Figure 6.3[19]. UV index computed from spectral measurements carried out by using the IMUK spectroradiometer: expected increment in the expanded uncertainty with the SZA (Cordero et al, 2008c for details).

---

[19] Figure 6.3 was adapted from Cordero et al, 2008c

Moreover, cosine error of the input optics should lead to an increment in the UV index uncertainty. Figure 6.3 shows the relative *expanded* uncertainty of the UV index, with different solar zenith angle. Because of the nearly Gaussian frequency distribution shown in figure 6.2, the expanded uncertainty was calculated from the standard uncertainty by applying a coverage factor equal to 2(see Cordero et al, 2008c for details).

7. Retrieval of Atmospheric parameters[20]

Ozone and aerosols lead to attenuation in the surface UV irradiance. These factors can be efficiently retrieved from ground-based measurements of the spectral direct UV irradiance.

As a final example, the newly proposed Monte Carlo-based technique (see section 3) was used to retrieve atmospheric parameters (ozone and aerosol information) from ground-based spectral measurements carried out by using the IMUK spectroradiometer.

In this case, the first step was to build a proper *retrieval model*; it implied sequentially comparing a measured spectrum with a large number of spectra, each of them computed by using randomly generated values of atmospheric parameters. Those generated values that led to a satisfactory match (a match was considered satisfactory when the difference between the compared spectra was within the uncertainty bounds) were taken to be *possible* values for the atmospheric parameters. The dispersion of these possible values allowed evaluating the retrieval uncertainty (by calculating the standard deviation).

7.1. Exploitation schema

A simple exploitation schema of spectral direct UV irradiance measurements is based on the recursive comparison of the ground-based measured spectra and model calculations. It allows retrieving both the ozone column and the Angström parameters α and β (and in turn the spectrally resolved aerosol optical depth AOD).

---

[20] Section 7 was adapted from Cordero et al, 2009

As shown in figure 7.1, such a procedure implies sequentially comparing the measured spectrum E_d and several spectral direct UV irradiances I_d, each of them calculated by using the UVSPEC model with values of α, β and O, randomly generated. The comparison involves calculating the ratio (E_d/I_d). If this ratio is not close to 1, the generated values of α, β and the ozone column are considered to be unlikely and they can therefore be discarded. Instead, the values that lead to ratios *reasonably close to* 1, can be considered to be likely and eventually taken to be the estimates of the α, β and O.

Although different criteria can be stated, as shown in figure 7.2, the ratio (E_d/I_d) was considered to be *reasonably close to* 1 (such that a good match between the measured and the calculated spectra was achieved) when the values of (E_d/I_d) lay within the bound specified by the involved uncertainties. Accordingly, the bound between the dotted lines in figure 7.2, was established by using the uncertainties of both I_d and E_d, estimated by following the procedure detailed in section 3 and 4, respectively.

If the generation of values of α, β and O, the direct UV irradiance calculation, and the subsequent comparison with the measured spectrum, are repeated a large number of times, several *likely* values of α, β and O, are expected to be found. This means that different values of α, β and O can lead to good matches between the measured and the calculated spectra. Instead of being a problem, as shown below, the dispersions of these values can be used to calculate the uncertainties of the aerosol parameters and the ozone column retrieved from the ground-based measurements.

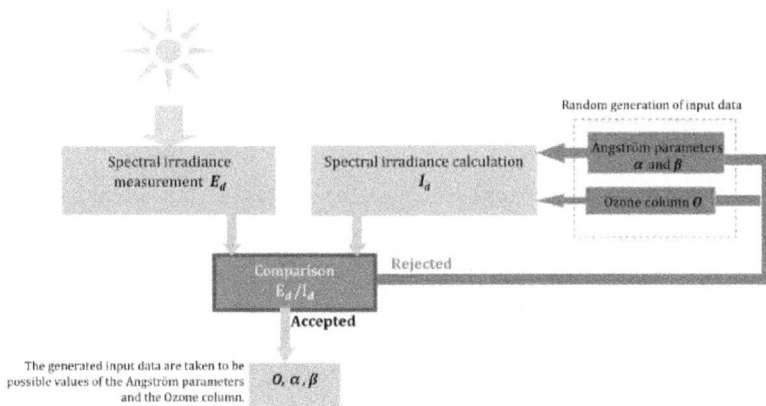

Figure 7.1[21]. Exploitation scheme. The comparison criterion is shown in figure 7.2.

7.2. Direct UV irradiance exploitation

The direct component of the UV irradiance, can be efficiently measured by using spectroradiometer systems with the input optics driven by a sun tracker. NDACC-certified mobile instruments, such as the IMUK spectroradiometer, are double monochromator-based instruments. Figure 7.3 shows the measurements of the direct spectral UV Irradiance performed at 30 min intervals on May 1st, 2007 (cloudless conditions) at Institut für Meterologie und Klimatologie, IMUK (Hannover, Germany) by using the IMUK spectroradiometer. The solar zenith angle at 12:30 h was about 40° and at 15:00 h was 57°.

Information concerning the aerosol properties (the Angström parameters α and β) as well as the ozone column O were retrieved from the measurements shown in figure 7.3 by applying the exploitation schema described in section 7.1. This implied setting an algorithm that sequentially compared the measurements and a relatively

---

[21] Figure 7.1 was adapted from Cordero et al, 2009

large number of spectra, each of them calculated by using values of α, β and O, randomly generated. The irradiance calculations were carried out by using the UVSPEC radiative transfer model (see section 4).

Figure 7.4 shows the results of the conducted Monte Carlo-based simulation process; this figure depicts the dispersion of the generated values of α, β and O, that allowed calculating a spectrum that matched reasonably well with the irradiance measured at 12:30 h. A match was considered to be acceptable when the differences between the compared spectra lay within the bound specified by the involved uncertainties.

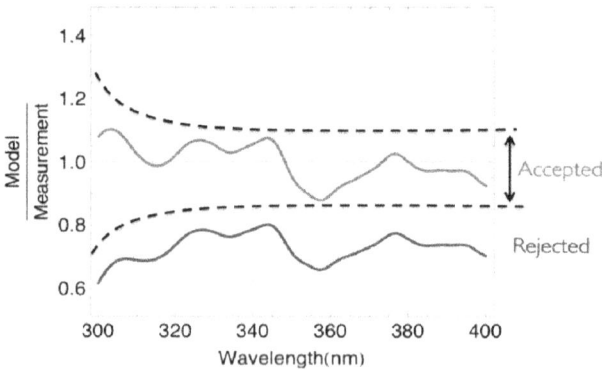

Figure 7.2[22]. Comparison between the calculated I_d and the measured E_d spectra. The bound between the dotted lines is defined by the uncertainty of both I_d and E_d.

The influence of the uncertainty sources affecting the measurements carried out by using the IMUK spectroradiometer has been comprehensively evaluated in section 5. Attending to these prior efforts, it was estimated that at solar zenith angles smaller than 70°, despite the variations due to wavelength shifts induced by the temperature during the measurements, the relative expanded uncertainties associated with

---

[22] Figure 7.2 was adapted from Cordero et al, 2009

measurements of the direct UV irradiance performed by using the IMUK spectroradiometer, are about 6% in the UV-A part of the spectrum; an increment is expected at wavelengths lower than 315 nm such that the expanded uncertainty of the UV-B irradiance at 290 nm wavelength can be up to 16%.

The direct component of the UV irradiance can be efficiently calculated by using radiative transfer models. These models allow solving by numerical means the equation of radiative transfer that governs the transfer of radiant energy in the atmosphere. The influence of the uncertainty sources affecting the direct UV irradiance values rendered by the UVSPEC model has already been characterized and compared with other systematic effects in section 4. This effort allowed estimating that, regardless of the solar zenith angles, under the cloudless conditions observed at the moment of the measurements, the relative standard uncertainties associated with UVSPEC calculations of the direct irradiance, are about 12% in the UV-A part of the spectrum; an increment is expected at UV-B wavelengths such that the standard uncertainty at 300 nm wavelength can be up to 20%.

Note that although the histograms in figure 7.4 were built up by using values of the Angström parameters and the ozone column that led to 18 good matches, the number of generated values of α, β and O (and in turn the number of comparisons performed during the simulation) was significantly greater. The number of simulations (and then the calculating time) that would allow building up histograms as those shown in figure 7.4, depends on the range, which the values of α, β and O can be randomly drawn from. This range can be set in the algorithm utilized to perform the simulation, by using some prior information that may be available. If information on the Angström parameters and the ozone column at the measurement location is available, the range of possible values of α, β and O can be restricted, constraining in turn the calculating time. The histograms in figure 7.4 were built up by performing a limited number of simulations that allowed keeping the calculating time shorter than 20 minutes even if a commercial PC was used.

(a)

(b)

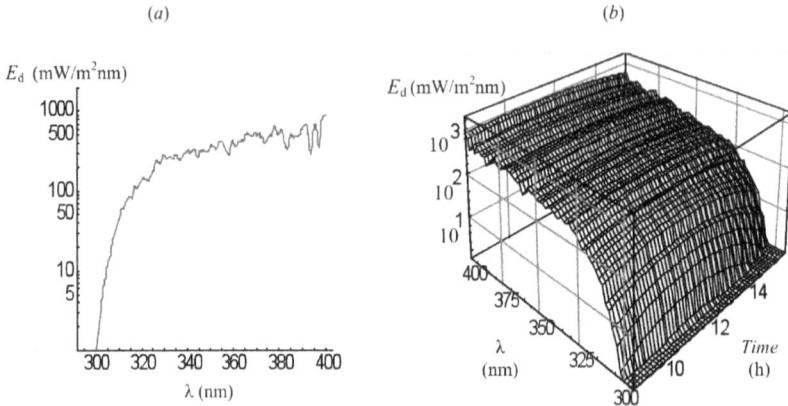

Figure 7.3[23]. Measurements performed at 30 min intervals on May 1st, 2007 (cloudless conditions) at Institut für Meterologie und Klimatologie, IMUK (Hannover, Germany) by using a spectroradiometer system of the Leibniz Universität Hannover. (a) Spectral direct UV irradiance measured at 12:30 h (the solar zenith angle was about 40°). (b) Spectral direct UV irradiance measured at the local time indicated in the plot; the solar zenith angle at 10:00 h was 40° and at 15:00 h was 57°.

---

[23] Figure 7.3 was quoted from Cordero et al, 2009

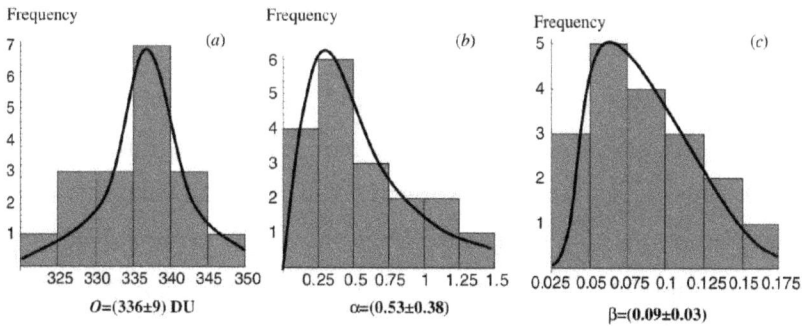

Figure 7.4[24]. Dispersion of possible values of (*a*) Ozone column O, (*b*) Angström parameter α, and *(c)* Angström parameter β, for the conditions observed at 12:30 h on May 1st, 2007 (cloudless conditions) at IMUK (Hannover, Germany).

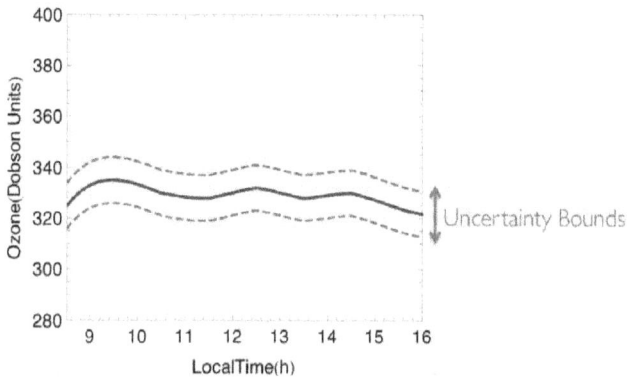

Figure 7.5[25]. The bold line indicates the estimates of the ozone column O, at different times on May 1st, 2007 (cloudless conditions) at Institut für Meterologie und Klimatologie, IMUK (Hannover, Germany). The dotted lines specify a bound within which O is expected to lie with a relatively high probability.

---

[24] Figure 7.4 was quoted from Cordero et al, 2009
[25] Figure 7.5 was adapted from Cordero et al, 2009

Since the all the values of α, β and O in figure 7.4 allowed calculating a spectrum that matched reasonably well with the measured irradiance, they can be considered to be likely. However the histograms in figure 7.4 indicate that some of these values are more probable. Although due to the limited number of performed simulations, the frequency distribution in figure 7.4 cannot allow identifying the probability density function (PDF) of α, β and O, they did allow calculating the estimate and the associated standard uncertainty of each of these parameters, by applying equations (3.1) and (3.2), respectively.

The estimates and the corresponding standard uncertainties of the dispersions shown in figure 7.4 are indicated close to each histogram. Note the relatively high uncertainty values associated with the estimate of α. This seems to be due to both the uncertainties in the particle size (which this Angström parameter stands for) and the limitations of the model in describing the spectral variations in the optical depth (after all, the Angström's law is only an approximation). Although the limitations of the Angström's law were not explicitly considered to be an error source in the uncertainty propagation, this was not necessary because the applied exploitation method implied the comparison of radiative transfer model outcomes and measured spectra. This comparison allowed implicitly including in the uncertainty evaluation the effect due to the Angström's law limitations, because, if these limitations do not allow properly following the spectral AOD variation, the dispersion of possible values of α should increase, leading in turn to greater uncertainties.

Nevertheless, these relatively high uncertainties were not a particularly surprising result considering that the values of α retrieved by applying related exploitation methods, are also expected to be highly uncertain (Eck et al, 1999; O'Neil et al,

2001]. However, as shown below, high uncertainties in the α parameter, does not necessarily lead to AODs particularly great uncertain.

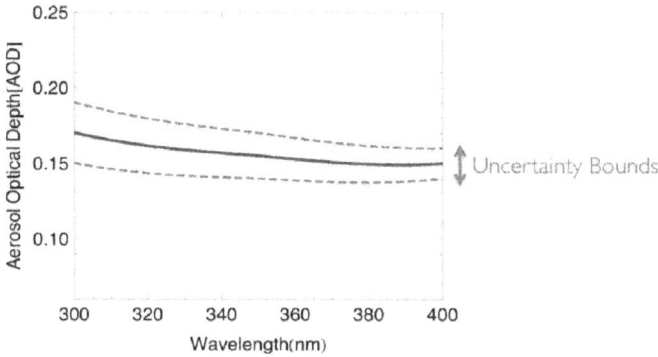

Figure 7.6[26]. The bold line indicates the spectrally resolved estimates of the aerosol optical depth AOD, for the conditions observed at 12:30 h on May 1st, 2007 (cloudless conditions) at IMUK (Hannover, Germany). The dotted lines define a bound within which the AOD is expected to lie with a relatively high probability.

7.3. Time series of total ozone column and aerosol optical depth

By applying the exploitation schema described in section 7.1 to each spectrum plotted in figure 7.3b, it was possible to follow the evolution of the aerosol properties (the Angström parameters α and β) as well as the ozone column O. Since these measurements were performed at 30 min intervals, both the estimates and corresponding standard uncertainties of the ozone column could be retrieved from these measurements at the same time interval.

The bold line in figure 7.5 was obtained by interpolating the estimates of the ozone column, retrieved each 30 minutes on May 1st, 2007 (cloudless conditions) at IMUK

---

[26] Figure 7.6 was adapted from Cordero et al, 2009

(Hannover, Germany). The corresponding uncertainties were used to build a bound (specified by the dotted lines in figure 7.5) within which O is expected to lie with a relatively high probability. Although similar plots can be built up by using the estimates and uncertainties of the Angström parameters, rather than the values of α and β, it can be more interesting to show the evaluation of the spectrally resolved AOD.

Since the aerosol optical depth can be calculated from the values of α and β, by applying the Angström's law ($AOD = \beta\lambda^{-\alpha}$), the computation of the AOD estimates retrieved from the ground-based measurements is straightforward. Instead, the evaluation of the uncertainty associated to these estimates, requires describing the uncertainty propagation through this equation. Since the Angström's law is a *nonlinear* model (the well known law of propagation of uncertainties is only recommended for linear or weakly nonlinear models), the uncertainty of the AOD from the uncertainties of α and β, was evaluated by using a Monte Carlo-based technique of uncertainty propagation. It implied applying the Angström's law and calculating the aerosol optical depth by using the pairs of Angström parameters $(\alpha_1, \beta_1, \alpha_2, \beta_2, ..., \alpha_N, \beta_N)$, that led to good matches between the measured and the calculated spectra. The recursive calculation rendered, at each wavelength, a set of indications $(AOD_1, ..., AOD_N)$. Then, the expected values and standard deviations of these indications can be calculated (by applying equations (3.1) and (3.2) respectively) and taken as being equal to the estimates and standard uncertainties of the AOD, at each wavelength.

The bold line in figure 7.6 indicates the spectrally resolved estimates of the AOD, retrieved from the direct spectral UV irradiances measured at 12:30 h (figure 7.3a). The corresponding uncertainties were used to build a bound (specified by the dotted lines in figure 7.6) within which AOD is expected to lie with a relatively high probability. The same procedure involving the measurements shown in figure 7.3b, allowed building up figure 7.7a; it shows the spectrally resolved estimates of the AOD, for the conditions observed on May 1st, 2007. The corresponding standard uncertainties are shown in figure 7.7b.

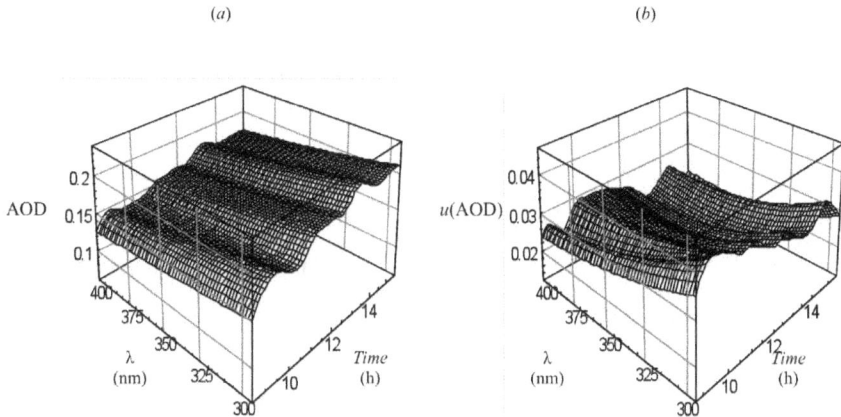

Figure 7.7[27]. (*a*) Spectrally resolved estimates of the aerosol optical depth AOD, for the conditions observed on May 1st, 2007 (cloudless conditions) at IMUK (Hannover, Germany). (*b*) Standard uncertainty of the optical depth values in (*a*).

It can be observed that although the AOD increased during the day (see figure 7.7a), the uncertainties of these values remains roughly constant (see figure 7.7b). Moreover, based on figures 7.6 and 7.7b, it is concluded that the uncertainty of AOD values retrieved from measurements of the spectral direct UV irradiance, tends to be greater at shorter wavelengths.

---

[27] Figure 7.7 was quoted from Cordero et al, 2009

8. Summary and Conclusions

An improved understanding of the global UV climate requires quality-ensured surface UV series. When developing an instrument, or in general terms, when developing a measurement model, quality assurance requires comparing the measurement with a reference. Since there is no underlying reference for spectral solar radiation measurements, testing the quality involves intercomparisons normally involving several instruments.

These comparisons require measurements all having stated uncertainties: the agreement is considered to be acceptable when the difference is within the uncertainty bounds. This means that quality assurance requires paying prior attention to the uncertainty evaluation.

Since conventional uncertainty propagation techniques cannot fully describe the nonlinear influence of uncertainty sources affecting UV spectroradiometry, a new Monte Carlo-based uncertainty propagation technique has been developed (see section 3).

By using the proposed Monte Carlo-based technique, the uncertainties of irradiances rendered by 1-D radiative transfer models and by spectroradiometers were evaluated (see section 4 and section 5, respectively). Since it allows comprehensively addressing the uncertainty propagation problem, the proposed Monte Carlo-based technique has the potential to become a useful tool for ensuring the quality of surface spectral UV measurements and in turn for assessing the performance of new instruments.

The quality-ensured UV spectra rendered by fully characterized instruments have different applications. A couple of them were also addressed: the computation of the

UV index (section 6), and the retrieval of atmospheric parameters (section 7); the proposed Monte Carlo-based method was applied in order to retrieve UV indexes and atmospheric parameters (both computed from ground-based measurements) as well as to evaluate their corresponding uncertainties.

8.1. Uncertainty Analysis of 1-D radiative transfer models

In section 4, an uncertainty analysis of the spectral UV irradiances (I) rendered by the UVSPEC model under cloudless sky conditions was carried out. In order to express the uncertainty of the output quantities (I) in terms of the standard uncertainties of the input quantities $(S_o, A, \theta, O, \omega, g, \alpha, \beta)$, the Monte Carlo-based uncertainty propagation technique, proposed in section 3, was used. It allowed considering the nonlinear effect on the output quantities due to some uncertainty sources affecting the input quantities.

The uncertainty propagation technique required first assigning Probability Density Functions (PDFs) to the input quantities needed to run the model: the extraterrestrial solar spectrum S_o, the ozone column O, the solar zenith angle θ, the surface albedo A, the asymmetry factor g, the single scattering albedo ω and the Angström parameters (α and β) used to stand for the spectral influence of aerosols. The assigned PDFs were scaled by using the uncertainty bounds attributed to the available values of $(S_o, A, \theta, O, \omega, g, \alpha, \beta)$. Next, the output quantity (I) was calculated a large number of times by using sets of data generated according to the assigned PDFs. Then, the standard deviations of the values of I generated by the large number of irradiance evaluations, were numerically computed and taken as the corresponding standard uncertainties: $u(I)$.

It was found that the main contributors to $u(I)$ in the UV-A part of the spectrum were the uncertainties attributed to the extraterrestrial spectrum S_o, the Angström

81

parameter β, and the single scattering albedo ω. The latter became particularly important in case of polluted air. On the other hand, the uncertainty of the irradiance in the UV-B part of the spectrum was significantly influenced by the uncertainty attributed to the ozone column datum O.

It was also found that the aerosol conditions strongly affected the irradiance uncertainties; the standard uncertainties of the global UV irradiances at 300 nm increased from about 9% under low aerosol conditions, up to about 20% in case of polluted air. Under conditions of great aerosol modulation, the influence of the uncertainties attributed to aerosol properties accounted for about 70% of $u(I)$ at wavelengths greater than 320 nm. It is concluded that the UV irradiance evaluation through radiative transfer models requires paying special attention to the assessment of the aerosols properties.

Although only the UVSPEC model was used in this work, the methodology applied to evaluate the uncertainty is general and it can be applied to any other model. Moreover, because a large fraction of the radiative transfer models are based on the same 1-D radiative transfer solver, the uncertainties associated with their outcomes should be in the same range of those reported above.

Note that the uncertainty estimations reported above accord well with Badosa et al, (2007) who compared measurements carried out by NDACC–certified instruments with radiative transfer model outcomes; for the case of the best input information available, Badosa et al, (2007) found an agreement within 6% for clean sites, and 10% for polluted sites. Agreement was acceptable because it was within the expected uncertainty bounds of the models.

The work on the performance of radiative transfer models allowed generating two papers:

82

Cordero RR. Seckmeyer G. Pissulla D. DaSilva L. Labbe F "Uncertainty evaluation of the spectral UV irradiance evaluated by using the UVSPEC Radiative Transfer Model" Optics Communications **276** (2007) 44-53

Cordero RR. Seckmeyer L. Labbe F "Evaluating the uncertainties of data rendered by computational models" Metrologia **44** (2007) L23-30

Section 4 is mostly based on these papers.

8.2. Uncertainty of double monochromator-based spectroradiometers

In section 5, an uncertainty analysis of the spectral irradiances E measured by using the IMUK spectroradiometer was carried out; this instrument complies with the requirements of the NDACC. The spectral measurements were performed during an international intercomparison campaign organized in the framework of the QASUME project.

The effects of temporal instabilities and nonlinearities in the signal were considered in the uncertainty evaluation. Moreover, the effect of the errors originated in the absolute calibration needed to carry out the measurements, were also explicitly considered in the uncertainty propagation. In order to express the uncertainty of the *output* quantity (the spectral irradiance) in terms of the uncertainties of the *input* quantities (all the experimental data obtained during the field measurements and the prior calibrations), the Monte Carlo-based uncertainty propagation technique proposed in section 3, was used again.

It was found an increment in relative uncertainty $u(E)/E$ at wavelengths shorter than 315 nm; this was attributed to the additive uncertainty affecting the measured signal, linked with eventual temporal offset variations. At solar zenith angles smaller than

30°, the uncertainty attributed only to the spectrum of the calibrating lamp accounted for about 60% of the UV-A uncertainty.

It should be noted that the *standard* uncertainty defines a bound within which the irradiance is expected to lie with a certain probability; because it was found that irradiances measured by the IMUK spectroradiometer can be described by using a normal frequency distribution, if the half-width of the bound is taken to be equal to the *standard* uncertainty, the irradiance should be in this interval with a probability of about 68%. The relative *expanded* uncertainty of the irradiance, $U(E)/E$, was calculated from the standard uncertainty $u(E)$ by applying a coverage factor equal to 2, such that $U(E)=2u(E)$. This coverage factor defines a bound within which the irradiance is expected to lie with a probability equal to about 95%. At solar zenith angles smaller than 30°, the relative expanded uncertainty at 300 nm was about 9%; it diminished with the wavelength such that the expanded uncertainty in the UV-A part of the spectrum was about 6%.

By comparison, these figures are significantly lower (by a factor 2) than those found in the case of UVSPEC-computed spectra under an unpolluted scenario (see section 4). However, they roughly agree with prior efforts: Bernhard and Seckmeyer (1999) applied the conventional uncertainty propagation technique to a double monochromator-based instrument; they assumed that the effect on the overall uncertainty, due to the involved error sources, was linear. This is only true when the equations (that link the output with the inputs) are linear. Based on that approximation, they computed the irradiance uncertainty by applying the LPU.

Instead, the Monte Carlo-based method applied in this work, implied recursively combining measurements and calibrations in the same way they usually are. This means that no approximations were made on how the involved error sources relate. This allowed fully accounting for the effect of both nonlinearities and correlations.

Despite their differences, the uncertainties estimated by using both techniques (conventional and Monte Carlo-based) roughly agreed in the case of double monochromator-based instruments. This was due to the strong influence of lamp-related errors on the overall uncertainty. The effects of other uncertainty sources were considerably smaller and then, although the Monte Carlo-based method was able to account for the influence of nonlinearities and correlations, their effect on the overall uncertainty was not significant.

Although only a double monochromator was used in this work, the methodology applied to evaluate the uncertainty is general and it agrees with recommendation of the ISO *Guide to the Expression of Uncertainty in Measurement*. Moreover, because the double monochromator systems of the NDACC network fulfill the same specifications and the rendered experimental data undergo the same quality control, the uncertainties associated with their outcomes should be similar to those reported above.

Note that although conventional uncertainty evaluation techniques allow defining a bound within which the irradiance is expected to lie, the proposed Monte Carlo-approach allows also estimating the probability of finding the irradiance within a certain bound; as pointed out above, if the expanded uncertainties are used to define the bound, the irradiance is expected to lie in that interval with a probability of 95%. However, the probabilies of finding the irradiance within a ±4% bound are only about 50%. This accords well with the work of Gröbner et al, (2006) who show results of an intercomparison of 25 European spectroradiometers relative to a transportable reference spectroradiometer. Half of the instruments agree with the reference spectroradiometer within a ±4% bound in the UV range.

The work on the uncertainty of double monochromator-based spectroradiometers allowed generating two papers:

Cordero RR. Seckmeyer G. Pissulla D. DaSilva L. Labbe F. "Uncertainty Evaluation of Spectral UV Irradiance Measurements" Meas. Sci. Technol. **19** (2008) 1-15

Cordero RR. Seckmeyer G. Labbe F. "Cosine error influence on ground-based spectral UV irradiance measurements" Metrologia **45** (2008) 406-414

Section 5 is mostly based on these papers.

8.3. UV index Uncertainty Analysis

In section 6, the Monte Carlo-based approach described in section 3 was applied to evaluate the uncertainty associated with the UV index. This is evaluated by calculating the integral in the range 250-400 nm of the spectral UV irradiance weighted by using the McKinlay-Diffey Erythema action spectrum. The spectral UV irradiance was approximated from a set of highly-resolved J points $\left(\lambda'_j, E_j\right)$, built up from experimental data measured by using the IMUK spectroradiometer (see section 5).

The measurements were performed during an international intercomparison campaign organized in the framework of the QASUME project.

As expected, it was found that the uncertainty of the UV index strongly depended on the uncertainty sources affecting the spectral irradiance measurements (the ordinates); the influences of the uncertainty sources affecting the values of the wavelengths (the abscissas) were relatively small. As a consequence, main contributors to the UV-B uncertainties became major influences on the UV index uncertainty.

In the case of the IMUK spectroradiometer, the driving factor determining the UV-B uncertainty was is dark signal. Although at wavelengths longer than 315 nm the contributions to the uncertainty due to the temporal variations in the dark signal were small, the increment in the relative uncertainty observed in the UV-B part of the spectrum, can be attributed to the additive variations in the measured signals due to that uncertainty source.

UV indexes with expanded uncertainties of about 6-8% can be computed from ground-based measurements carried out by using the IMUK spectroradiometer. In the case of solar zenith angles greater than 30 degrees, the cosine error of the input optics should lead to an increment in the UV index uncertainty. That increment depends of course on the angular response of the input optics but for state-of-the-art input optics, expanded uncertainties lower than 10% are always expected.

The work on the uncertainty of UV indexes computed from ground-based spectral measurements allowed generating the paper:

Cordero RR. Seckmeyer G. Pissulla D. Labbe F. "Uncertainty of experimental integrals: application to the UV index calculation" Metrologia **45** (2008) 1-10

Section 6 is mostly based on this paper.

8.4. Retrieval of aerosol parameters

In section 7, the ozone column and the aerosol properties (AOD, α and β) were retrieved from direct UV irradiances, by using a Monte Carlo-based retrieval method. The retrieval method was based on comparing the measured direct UV spectra with calculations carried out by using the UVSPEC radiative transfer model.

The spectral measurements were performed by using the IMUK spectroradiometer under cloudless conditions in Hannover (Germany) during an intercomparison campaign organized in the framework of the SCOUT project.

The propagation of the uncertainty through the retrieving process was described by using a Monte Carlo-based technique, which implied sequentially comparing the ground-based measurements and a large number of spectra, each of them calculated by using randomly generated values of α, β and O. Some of the generated values of α, β and O, led to a calculated spectrum that matched reasonably well with the measured irradiance. A match was considered to be acceptable when the differences between the compared spectra lay within the bound specified by the combined effect of the uncertainties affecting both measurements and calculations.

Afterwards, it was possible to evaluate the estimates and corresponding uncertainties of the ozone column as well as the Angström parameters by calculating the expected value and the standard deviation of the set of N values of α, β and O that led to acceptable matches. The pairs of Angström parameters $(\alpha_1, \beta_1, \alpha_2, \beta_2, ..., \alpha_N, \beta_N)$ that led to those acceptable matches were in turn used to evaluate the AOD uncertainty by recursively applying the Angström's law. At each wavelength, the expected values and standard deviations of the set of indications $(AOD_1, ..., AOD_N)$ rendered by the Angström's law application, allowed calculating the estimates and standard uncertainties of the aerosol optical depth.

It was found that, despite the variations in the AOD and the ozone values retrieved from direct UV spectra, the uncertainty associated with these values remained roughly constant. Moreover, the AOD uncertainty was consistently found to be greater at shorter wavelengths. Expanded uncertainties of about 8% for the ozone column, and of about 22% for AOD retrievals, were found when exploiting direct UV irradiances.

Because it allows retrieving both estimates and uncertainties, the applied Monte Carlo-based exploitation technique of spectral UV measurements, renders a bound within which the retrieved parameter (either the ozone column or an aerosol property) is expected to lie with a relatively high probability. Attending to the significant influence of the uncertainty sources involved in any retrieving process, this seems to be an advantage when compared with techniques that yield singles values.

The work on the retrieval of ozone and aerosol parameters from ground-based measurements allowed generating the paper:

Cordero RR, Seckmeyer G, Pissulla D, Labbe F, "Exploitation of Spectral Direct UV Irradiance Measurements" Metrologia **46** (2009) 19-25

Section 7 is mostly based on this paper.

8.5. Final remarks

In cases where inputs lead to nonlinear effects on the output, reliable uncertainty evaluations require nonconventional techniques. Accordingly, a Monte Carlo-based technique was proposed in order to fully describe the uncertainty propagation when gathering spectral UV data.

The proposed technique allowed comprehensively evaluating the uncertainty of surface UV spectra, computed on the one hand by using 1-D radiative transfer models (see section 4), and measured on the other hand by using spectroradiometers (see section 5).

Surface spectral UV data are normally exploited in order to compute UV indexes and also to retrieve atmospheric parameters (ozone column or aerosol load). Although several existing techniques enable exploiting UV spectra, they do not render information on the retrieval uncertainties. As shown below, the proposed Monte Carlo-based technique also enabled fully describing the uncertainty propagation through the retrieving process; it allowed expressing the uncertainty of retrievals (UV indexes and atmospheric parameters) in terms of the uncertainties of all the experimental data used to build up the exploited UV spectra. By doing so, the proposed technique became ultimately an exploitation tool; it enabled computing UV indexes from global UV irradiances (see section 6), and retrieving ozone and AOD values from direct UV irradiances (see section 7), all having stated uncertainties.

The examples shown above exposed the potential of the new Monte Carlo-based approach. The technique allows comprehensively describing the uncertainty propagation through any measuring method or any retrieving process. Therefore, it can become a useful tool for exploiting spectral UV measurements and for ensuring their quality.

This work allowed so far generating 8 peer-reviewed manuscripts already published by Metrologia, Measurement Science and Technology, Optics Communications and Photochemical & Photobiological Sciences. The text above is mostly based on those papers.

References

Anderson G P, Clough S A, Kneizys F X, Chetwynd J H and Shettle E P, 1986 AFGL Atmospheric Constituent Profiles (0-120 km), AFGL-TR-86-0110, AFGL (OPI), Hanscom AFB, MA 01736

Ansko I, Eerme K, Latt S, Noorma M, and Veismann U, 2008, Study of suitability of AvaSpec array spectrometer for solar UV field measurements, *Atmos. Chem. Phys.* 8, 3247–53,

Autier P, Gandini S, 2007, Vitamin D Supplementation and Total Mortality; A Meta-analysis of Randomized Controlled Trials, *Arch. Intern. Med.* 167(16), 10-15

Badosa J, McKenzie R L, Kotkamp M, Calbó J, González JA, Johnston PV, O'Neill M, and Anderson D J, 2007, Towards closure between measured and modelled UV under clear skies at four diverse sites, *Atmos. Chem. Phys.* 7, 2817-37

Bais A F, Gardiner B G, Slaper H, Blumthaler M, Bernhard G, McKenzie R, Webb A R, Seckmeyer G, Kjeldstad B, Koskela T, Kirsch P J, Groebner J, Kerr J B, Kazadzis S, Leszczynski K, Wardle D, Josefsson W, Brogniez C, Gillotay D, Reinen H, Weihs P, Svenoe T, Eriksen P, Kuik F, and Redondas A, 2001, The SUSPEN intercomparison of ultraviolet spectroradiometers, *J. Geophys. Res.* 106, 12509–26

Bais A F, 1997, Spectroradiometers: operational errors and uncertainties, in: Solar Ultraviolet Radiation. Modelling, Measurements and Effects, edited by C Zerefos and A Bais (Springer), 52, 165–73

Bernhard G, Booth C R, and Ehramjian J C, 2005, UV climatology at Palmer Station, Antarctica, in: Ultraviolet Ground- and Space-based Measurements, Models, and Effects V, edited by G Bernhard, J R Slusser, J R Herman and W Gao, Proceedings of SPIE, 588607-1 - 588607-12.

Bernhard G and Seckmeyer G, 1999, Uncertainty of measurements of spectral solar UV irradiance, *J. Geophys. Res.* 104, 14321–45

Bernhard G, Booth C R, and Ehramjian J C, 2010, Climatology of Ultraviolet Radiation at High Latitudes Derived from Measurements of the National Science Foundation's Ultraviolet Spectral Irradiance Monitoring Network, in: *UV Radiation in Global Climate Change: Measurements, Modeling and Effects on Ecosystems*, edited by W Gao, D L Schmoldt and J R Slusser (Springer-Verlag and Tsinghua University Press) ISBN: 978-3-642-03312-4

Bogh M K B, Schmedes A V, Philipsen P A, Thieden E, Wulf H C, 2011, Vitamin D production depends on ultraviolet-B dose but not on dose rate: A randomized controlled trial, *Experimental Dermatology* 20 (1), 14–8,

Buchard, V, Brogniez C, Auriol F, Bonnel B, Lenoble J, Tanskanen A, Bojkov B, and Veefkind P, 2008, Comparison of OMI ozone and UV irradiance data with ground-based measurements at two French sites, *Atmos. Chem. Phys.* 8, 4517–28

Cachorro V E, Vergaz R, Martin M J, de Frutos A M, Vilaplana J M and de la Morena B, 2002, Measurements and estimations of the columnar optical depth of tropospheric aerosols in the UV spectral region, *Ann. Geophys.* 20, 565-74

Cede A, Herman J, Richter A, Krotkov N and Burrows J, 2006, Measurements of nitrogen dioxide total column amounts using a Brewer double spectrophotometer in direct sun mode, *J. Geophys. Res.* 111

Coleman A, Sarkany R and Walker S, 2008, Clinical ultraviolet dosimetry with a CCD monochromator array spectroradiometer, *Phys. Med. Biol.* 53, 5239–55

Cordero R R and Roth P, 2004, Assigning Probability Density Functions in a Context of Information Shortage *Metrologia* 41, L22–L25

Cordero R R and Roth P. 2005, Revisiting the problem of the evaluation of the uncertainty associated with a single measurement, *Metrologia* 42, L15–L19

Cordero R R, Seckmeyer G and Labbe F, 2006, Effect of the resolution on the uncertainty evaluation, *Metrologia* 43, L33–L38

Cordero R R, Seckmeyer G, Pissulla D, DaSilva L and Labbe F, 2007a, Uncertainty evaluation of the spectral UV irradiance evaluated by using the UVSPEC Radiative Transfer Model, *Optics Communications* 276, 44-53

Cordero R R, Seckmeyer G and Labbe F, 2007b, Evaluating the uncertainties of data rendered by computational models, *Metrologia* 44, L23-30

Cordero R R, Seckmeyer G, Pissulla D, DaSilva L and Labbe F, 2008a, Uncertainty Evaluation of Spectral UV Irradiance Measurements, *Meas. Sci. Technol.* 19, 1-15

Cordero R R, Seckmeyer G, Pissulla D and Labbe F, 2008b, Uncertainty of experimental integrals: application to the UV index calculation, *Metrologia* 45, 1-10

Cordero R R, Seckmeyer G and Labbe F, 2008c, Cosine error influence on ground-based spectral UV irradiance measurements, *Metrologia* 45, 406-414

Cordero R R, Seckmeyer G, Pissulla D and Labbe F, 2009, Exploitation of Spectral Direct UV Irradiance Measurements, *Metrologia* 46, 19-25

Dahlback A and Stamnes K, 1991, A new spherical model for computing the radiation field available for photolysis and heating at twilight, *Planet. Space Sci.* 39, 671–683

Davis A and Marshak A, 2010, Solar radiation transport in the cloudy atmosphere: a 3D perspective on observations and climate impacts, *Rep. Prog. Phys.* 73, 026801

Dubovik O, Smirnov A, Holben BN, King MD, Kaufman YJ, Eck TF, and Slutsker I, 2000, Accuracy assessment of aerosol optical properties retrieval from AERONET sun and sky radiance measurements, *J. Geophys. Res.* 105, 9791–806.

Dubovik O, Holben B, Eck T F, Smirnov A, Kaufman Y J, King M D, Tanre D and Slutsker I, 2000, Variability of Absorption and Optical Properties of Key Aerosol Types Observed in Worldwide Locations, *J. Atmos. Sci.* 59, 590-608

Eck T F, Holben B N, Reid J S, Dubovik O, Smirnov A, O'Neill N T, Slutsker I and Kinne S, 1999, Wavelength dependence of the optical depth of biomass burning, urban and desert aerosols, *J. Geophys. Res.* 104, 31,333–31,349

Foyo-Moreno I, Alados I, Olmo F J and Alados-Arboledas L, 2003, The influence of cloudiness on UV global irradiance (295–385 nm), *Agricultural and Forest Meteorology* 120, 101-111

Gary E T and Stamnes K, 1999, Radiative transfer in the atmosphere and ocean, (Cambridge University Press) 517 p.

Gröbner J, Blumthaler M, Kazadzis S, Bais A, Webb A, Schreder J, Seckmeyer G, and Rembges D, 2006, Quality assurance of spectral solar UV measurements: result from 25 UV monitoring sites in Europe, 2002 to 2004, *Metrologia* 43, S66-S71.

Gröbner J, Albold A, Blumthaler M, Cabot T, de la Casinière A, Lenoble J, Martin T, Masserot D, Müller M, Philipona R, Pichler T, Pougatch E, Rengarajan G, Schmucki D, Seckmeyer G, Sergent C, Tour ML, and Weihs P, 2000, The variability of spectral solar ultraviolet irradiance in an Alpine environment, *J. Geophys. Res*. 105, 26991–7003

Gueymard CA, 2004, The sun's total and spectral irradiance for solar energy applications and solar radiation models, *Sol. Energy* 76, 423-453

Gueymard C A, Myers D and Emery K, 2002, Proposed reference irradiance spectra for solar energy systems testing, *Sol. Energy* 73 (6), 443-467

Gueymard C A, 2006, Reference solar spectra: Their evolution, standardization issues, and comparison to recent measurements, *Adv. Space Res*. 37, 323–340

Hendrick F, Pommereau J-P, Goutail F, Evans R D, Ionov D, Pazmino A, Kyrö E, Held G. Eriksen P, Dorokhov V, Gil M and Van Roozendael M, 2011, NDACC/SAOZ UV-visible total ozone measurements: improved retrieval and comparison with correlative ground-based and satellite observations, *Atmos. Chem. Phys*. 11, 5975-5995.

Holben B N, Eck T I, Slutsker I, Tanre D, Buis J P, Setzer A, Vermote E, Reagan JA, Kaufman Y, Nakajima T, Lavenu F, Jankowiak I and Smirnov A, 1998, AERONET-A Federated Instrument Network and Data Archive for Aerosol Characterization, *Remote Sens. Enviro*. 66, 1–16

Holick M F, 2008, Sunlight UV-Radiation, Vitamin D and Skin Cancer: How Much Sunlight Do We Need? *Advances in Experimental Medicine and Biology* 624, 1-15

Huber M, Blumthaler M, Ambach W and Staehelin J, 1995, Total atmospheric ozone determined from spectral measurements of direct solar UV irradiance *Geophys. Res. Lett.* 22, 53–56

Ialongo I, Casale G R, and Siani A M, 2008, Comparison of total ozone and erythemal UV data from OMI with ground-based measurements at Rome station, *Atmos. Chem. Phys.* 8, 3283-3289

Ialongo, I, Buchard V, Brogniez C, Casale G R, and Siani A M, 2009, Aerosol Single Scattering Albedo retrieval in the UV range: an application to OMI satellite validation, *Atmos. Chem. Phys. Discuss.* 9, 19009-19033

ISO, 1993, Guide to the Expression of Uncertainty in Measurement (Geneva: ISO)

ISO, 2004, Guide to the Expression of Uncertainty in Measurement, Supplement 1: Numerical Methods for the Propagation of Distributions (Geneva: ISO)

Jacovides C P, Steven M D, Asimakopoulos D N, 2000, Spectral Solar Irradiance And Some Optical Properties For Various Polluted Atmospheres, *Solar Energy*, 69 (3), 215-227

Jaekel E, Wendisch M, Blumthaler M, Schmitt R and Webb A, 2007, A CCD spectroradiometer for ultraviolet actinic radiation measurements, *J. Atmos. Oceanic Technol.* 24(3), 449–462

Janouch M, and Metelka L, 2007, Modeling UV spectra with help of neural network, p.207-2008, Proceedings UV Conference "One century of UV Radiation Research", Davos, Switzerland. Eds. J.Grobner

Junk J, Feister U and Helbig A, 2007, Reconstruction of daily solar UV radiation from 1893 to 2002 in Potsdam, Germany, *International Journal of Biometeorology* 51, 505-12

Kato S and Marshak A, 2009, Solar zenith and viewing geometry-dependent errors in satellite retrieved cloud optical thickness: Marine stratocumulus case, *J. Geophys. Res.* 114, D01202

Kazadzis S, Bais A, Balis D, Kouremeti N, Zempila M, Arola A, Giannakaki E, Amiridis V and Kazantzidis A, 2009a, Spatial and temporal UV irradiance and aerosol variability within the area of an OMI satellite pixel. *Atmos. Chem. Phys.* 9, 4593–601

Kazadzis S, Bais A, Arola A, Krotkov N, Kouremeti N, and Meleti C, 2009b, Ozone Monitoring Instrument spectral UV irradiance products: comparison with ground based measurements at an urban environment, *Atmos. Chem. Phys.* 9, 585–94

Kylling A, Persen T, Mayer B, and Svenoe T, 2000, Determination of an effective spectral surface albedo from ground based global and direct UV irradiance measurements, *J. Geophys. Res.* 105 (D4), 4949-59

Kouremeti N, Bais A, Kazadzis S, Blumthaler M, and Schmitt R, 2008, Charge-coupled device spectrograph for direct solar irradiance and sky radiance measurements, *Appl. Opt.* 47, 1594-1607

Kreuter A and Blumthaler M, 2009, Stray light correction for solar measurements using array spectrometers, *Rev. Sci. Instrum.* 80, 096108

Krzyscin J W, Jarosawski J and Sobolewski P S, 2003, Effects of clouds on the surface erythemal UV-B irradiance at northern midlatitudes: estimation from the observations taken at Belsk, Poland (1999–2001), *Journal of Atmospheric and Solar-Terrestrial Physics* 65(4), 457-67

Lenoble J, 1993, Atmospheric Radiative Transfer (Hampton: A. Deepak Publishing)

Lira I, 2002, Evaluating the Uncertainty of Measurement:Fundamentals and Practical Guidance (Bristol: Institute of Physics Publishing)

Manney G L, Santee M L, Rex M, Livesey N J, Pitts M C, Veefkind P, Nash E R, Wohltmann I, et al, Unprecedented Arctic ozone loss in 2011, *Nature* 478, 469–75

Martin T, Gardiner B, and Seckmeyer G, 2001, Uncertainties in satellite-derived estimates of surface UV doses, *J. Geophys. Res.* 105, 27005–12

Matthijsen J, Slaper H, Reinen H A G M and Velders G J M, 2000, Reduction of solar UV by clouds: A remote sensing approach compared with ground based measurements, *J. Geophys. Res.* 105, 5069-80.

Mayer B, 1999, I3RC phase 1 results from the MYSTIC Monte Carlo model, in Intercomparison of Three-Dimensional Radiation Codes: Abstracts of the First and Second International Workshops, pp. 49– 54, Univ. of Ariz. Press, Tucson.

Mayer B, 2000, I3RC phase 2 results from the MYSTIC Monte Carlo model, in Intercomparison of Three-Dimensional Radiation Codes: Abstracts of the First and Second International Workshops, pp. 107– 108, Univ. of Ariz. Press, Tucson.

Mayer B and Kylling A, 2005, Technical note: The libRadtran software package for radiative transfer calculations - description and examples of use, *Atmos. Chem. Phys.* 5 1319–81

Mayer B and Seckmeyer G, 1998, Retrieving Ozone Columns from Spectral Direct and Global UV Irradiance Measurements, proceedings of the Quadrennial Ozone Symposium L'Aquila ed. Rumen D.Bojkov and Guido Visconti 935-938

Mayer B, Seckmeyer G and Kylling A, 1997, Systematic longterm comparison of spectral UV measurements and UVSPEC modeling results *J. Geophys. Res.* 102(D7), 8755–67

McKenzie R, Johnston P and Seckmeyer G, 1997, UV spectroradiometry in the network for the detection of stratospheric change (NDSC) ed C Zerefos and A Bais Solar Ultraviolet Radiation, Modelling, Measurements and Effects (Berlin: Springer) pp 279–87

McKenzie R L, Seckmeyer G, Bais A F, Kerr J B, and Madronich S, 2001, Satellite-retrievals of erythemal UV dose compared with ground-based measurements at Northern and Southern mid-latitude, *J. Geophys. Res.* 106 D20, 24051-62

McKinlay A F, Diffey B L, 1987, A reference action spectrum for ultra-violet induced erythema in human skin. In *Human Exposure to Ultraviolet Radiation: Risks and Regulations*. International Congress Series. Passchier WF, Bosnjakovich BFM, Eds. (Elsevier: Amsterdam) p 83-87

O'Neil N T, Eck T F, Holben B N, Smirnov A, Dubovik O, and Royer A, 2001, Bimodal size distribution influences on the variation of Angström derivatives in spectral and optical depth space *J. Geophys. Res.* 106, 9787–806

Riechelmann S, 2008, Messung von spektraler Bestrahlungsstärke und Strahldichte mit CCD-Array Geräten (in German), Diplomarbeit im Fach Meteorologie, Institut für Meteorologie und Klimatologie Gottfried Wilhelm Leibniz Universität Hannover

Satheesh S K, Srinivasan J, Vinoj V, Chandra S, 2006, New Directions: How representative are aerosol radiative impact assessments? *Atmospheric Environment* 40(16), 3008–10

Seckmeyer G, Pissulla D, Glandorf M, Henriques D, Johnsen B, Webb A, Siani A M, Bais A, Kjeldstad B, Brogniez C, Lenoble J, Gardiner B, Kirsch P, Koskela T, Kaurola J, Uhlmann B, Slaper H, den Outer P, Janouch M, Werle P, Grobner J, Mayer B, de la Casiniere A, Simic S and Carvalho F, 2008, Variability of UV Irradiance in Europe, *Photochemistry and Photobiology* 84, 172–179

Seckmeyer G, Bais A, Bernhard G, Blumthaler M, Booth C R, Disterhoft P, Eriksen P, McKenzie R L, Miyauchi M and Roy C, 2001, Part 1: Spectral instruments Instruments to Measure Solar Ultraviolet Radiation WMO-GAW No. 125 (Geneva, Switzerland: World Meteorological Organization)

Seckmeyer G and McKenzie R, 1992, Increased ultraviolet radiation in New Zealand (45°S) relative to Germany (48°N), *Nature* 359, 135-137

Seckmeyer G and Bernhard G, 1993, Cosine error correction of spectral UV irradiances *Atmospheric Radiation* ed K Stamnes *Proc. SPIE—The International Society for Optical Engineering* vol 2049 pp 140–51

Schwander H, Koepke P and Ruggaber A, 1997, Uncertainties in modelled UV irradiances due to limited accuracy and availability of input data, *J. Geophys. Res.* 102(D8), 9419–29

Shettle E P, 1989, Models of aerosols, clouds and precipitation for atmospheric propagation studies, in "Atmospheric propagation in the uv, visible, ir and mm-region and related system aspects", AGARD Conference Proceedings (454).

Slaper H, Reinen HA, Blumthaler M, Huber M and Kuik F, 1995, Comparing ground-level spectrally resolved solar UV measurements using various instruments: A technique resolving effects of wavelength shift and slit width. *Geophysical Research Letters* 22, 2721-4.

Slaper H, Velders G J M, Daniel J S, de Gruijl F R, van der Leun J C, 1996, Estimates of ozone depletion and skin cancer incidence to examine the Vienna Convention achievements. *Nature* 384, 256-8

Smolskaia I, 2001, Effect of inhomogeneous surface albedo on UV radiation in the Antarctic environment, PhD thesis, IASOS, University of Tasmania.

Smolskaia I, Wuttke S, Seckmeyer G and Michael K, 2006, Influence of surface reflectivity on radiation in the Antarctic environment, SPIE Proceedings, Vol. 6362.

Solomon S, Portmann R W and Thompson D W J, 2007, Contrasts between antarctic and arctic ozone depletion. Proc. Natl. Acad. Sci. U. S. A., 104 (2) 445-9

Stamnes K, Tsay S C, Wiscombe W and Jayaweera K, 1988, Numerically stable algorithm for discrete-ordinate-method radiative transfer in multiple scattering and emitting layered media, *Appl. Optics* 27, 12-15

Tanskanen A, Krotkov N A, Herman J R, and Arola A, 2006, Surface ultraviolet irradiance from OMI, *IEEE T. Geosci. Remote* 44(5), 1267–71

Tanskanen A, Lindfors A, Määttä A, Krotkov N, Herman J, Kaurola J, Koskela T, Lakkala K, Fioletov V, Bernhard G, McKenzie R, Kondo Y, O'Neill M, Slaper H, den Outer P, Bais A F and Tamminen J, 2007a, Validation of daily erythemal doses from Ozone Monitoring Instrument with ground-based UV measurement data, *J. Geophys. Res. 112*, D24S44

Tanskane A and Manninen T, 2007b, Effective UV surface albedo of seasonally snow-covered lands, *Atmos. Chem. Phys.* 7, 2759–64

Tarasick D W, Fioletov V E, Wardle J B, Kerr J B, McArthur K J B and McLinden C S, 2003, Climatology and trends of surface UV radiation, *Atmos. Ocean* 41, 121–38

Tevini M, 1993, UV-B Radiation and Ozone Depletion: Effect on Humans, Animals, Plants, Microorganisms and Materials (New York: Lewis)

Thuillier G, Hersé M, Labs D, Foujols T, Peetermans W, Gillotay D, Simon P C and Mandel H, 2003, The Solar Spectral Irradiance From 200 To 2400 nm As Measured By The Solspec Spectrometer From The Atlas and Eureca Missions, *Sol. Phys.* 214, 1–22

Uchino O, Bojkov RD, Balis DS, Akagi K, Hayashi M and Kajihara R, 1999, Essential characteristics of the Antarctic-spring ozone decline: Update to 1998, *Geophysical Research Letters* 26(10), 1377-80

Udelhofen P M, Gies R and Roy C, 1999, Surface UV radiation over Australia, 1979-1992: Effects of ozone and cloud cover changes on variations of UV radiation, *J. Geophys. Res.* 104, 19135-59

Verdebout J, 2004a, A European satellite-derived UV climatology available for impact studies, *Radiat. Prot. Dosim.* 111(4), 407-11

Verdebout J, 2004b, A satellite-derived UV radiation climatology over Europe to support impact studies, *Arct. Antarct. Alp. Res.* 36(3), 357-63

Weihs P and Webb A R, 1997a, Accuracy of spectral UV model calculations 2. Comparison of UV calculations with measurements, *J. Geophys. Res.* 102(D1), 1551–60

Weihs P and Webb A R, 1997b, Accuracy of spectral UV model calculations 1. Consideration of uncertainties in input parameters, *J. Geophys. Res.* 102(D1) 1541–50

Weihs P, Simic S, Laube W, Mikielewicz W and Rengarajan G, 1999, Albedo influences on surface UV irradiance at the Sonnblick High Mountain Observatory (3106 m altitude), *J. Appl. Meteorol.* 38(11), 1599–610.

Weihs P, Blumthaler M, Rieder H E, Kreuter A, Simic S, Laube W, Schmalwieser A W, Wagner J E, and TanskanenA, 2008, Measurements of UV irradiance within the area of one satellite pixel, *Atmos. Chem. Phys.* 8, 5615–26

WMO(World Meteorological Organization), WMO Report on the WMO–WHO meeting of experts on standardization of UV Indices and their dissemination to the public, WMO GAW 127, 1997

WMO (World Meteorological Organization), *Scientific Assessment of Ozone Depletion: 2006,* Global Ozone Research and Monitoring Project—Report No. 50, Geneva, Switzerland, 2007

WMO (World Meteorological Organization), *Scientific Assessment of Ozone Depletion: 2010,* Global Ozone Research and Monitoring Project—Report No. 52, Geneva, Switzerland, 2011

Wuttke S, Verdebout J and Seckmeyer G, 2003, An improved algorithm for satellite-derived UV radiation, *Photochem. & Photobiol.* 77(1), 52-7

Wuttke S, Seckmeyer G, Bernhard G, Ehramjian J, McKenzie R, Johnston P and O'Neil N, 2006, New spectroradiometers complying with the NDSC standards, *J. Atmospheric Oceanic Technol.* 23, 241–51

Ylianttila L, Visuri R, Huurto L and Jokela K, 2005, Evaluation of a Single-monochromator Diode Array Spectroradiometer for Sunbed UV-radiation Measurements, *Photochemistry and Photobiology* 81, 333-341

Zerefos C S, 2002, Long-term ozone and UV variations at Thessaloniki, Greece, *Phys. Chem. Earth* 27(6–8), 455–60

Zinner T, Mayer B and Schroder M, 2006, Determination of three-dimensional cloud structures from high-resolution radiance data, *J. Geophys. Res.* 111, D08204

Zong Y, Brown S W, Johnson B C, Lykke K R, and Ohno Y, 2006, Simple spectral stray light correction method for array, *Applied Optics* 45(6), 1111-9

Symbols

Abbreviations

AOD   Aerosol optical depth

CCD   Charge coupled device

FWHM   Full width at half maximum

IMUK   Institut für Meteorologie und Klimatologie

LPU   Law of propagation of uncertainties

LPD   Law of propagation of distributions

NDACC   Network for the Detection of Atmospheric Composition Change

NDSC   Network for the Detection of Stratospheric Change

NIST   National Institute Standards and Technology

PDF   Probability density function

PME   Principle of Maximum Entropy

QASUME Quality Assurance of Spectral Ultraviolet Measurements in Europe

SZA   Solar zenith angle

TOMS   Total Ozone Mapping Spectrometer

UV	Ultraviolet
UVA	320-400 nm wavelength radiation
UVB	250-320 nm wavelength radiation
UVI	UV index
UVSPEC	Radiative transfer model
WMO	World Meteorological Organization

Latin Symbols

A	Albedo
B	Correction factor for cosine error influence
b	Cosine error
d	Error bound
g	Asymmetry factor
E	Measured global irradiance
E_d	Measured direct irradiance
E_o	Underlying biologically weighted irradiance

E_w	Biologically weighted Irradiance
I	Calculated global irradiance
I_d	Calculated direct irradiance
I_n	UV index
L	Radiance distribution
O	Ozone column
P	Input quantity
$\bar{q}$	Sample estimated value
Q	Output quantity
p	Measured Input quantity
Q	Measured Output quantity
r	Resposivity
s	Sample standard deviation
S_o	Extraterrestrial Spectrum
S	Indicated Signal values
S'	Corrected Signal values

u	Standard Uncertainty
v	Multiplicative Correction factor for the signal
w	Additive Correction factor for the signal
U	Extended Uncertainty
z	Additive Correction factor for the wavelength

Greek Symbols

α	Angström's parameters
β	
δ	Resolution
$\Delta\theta$	Deviation from ideal cosine response
λ	Wavelength indicated by a spectroradiometer
λ'	Corrected Wavelength
σ_e	Extinction coefficient
σ_a	Absorption coefficient

σ_s	Scattering coefficient
θ	Solar zenith angle
ϑ	Azimuth angle
ω	Single scattering albedo

www.ingramcontent.com/pod-product-compliance
Lightning Source LLC
Chambersburg PA
CBHW021116210326
41598CB00017B/1463